www.EffortlessMath.com

... So Much More Online!

- ✓ FREE Math lessons
- ✓ More Math learning books!
- ✓ Mathematics Worksheets
- ✓ Online Math Tutors

Need a PDF version of this book?

Visit www.EffortlessMath.com

Or send email to: info@EffortlessMath.com

SSAT Upper Level Mathematics Prep 2019

A Comprehensive Review and Ultimate Guide to the SSAT Upper Level Math Test

By

Reza Nazari & Ava Ross

All inquiries should be addressed to:

info@effortlessMath.com

www.EffortlessMath.com

ISBN-13: 978-1-970036-04-6

ISBN-10: 1-970036-04-4

Published by:

Effortless Math Education

www.EffortlessMath.com

Description

SSAT Upper Level Mathematics Prep 2019 provides students with the confidence and math skills they need to succeed on the SSAT Upper Level Math, building a solid foundation of basic Math topics with abundant exercises for each topic. It is designed to address the needs of SSAT UPPER LEVEL test takers who must have a working knowledge of basic Math.

This comprehensive book with over 2,500 sample questions and 2 complete SSAT Upper Level tests is all you need to fully prepare for the SSAT Upper Level Math. It will help you learn everything you need to ace the math section of the SSAT Upper Level.

Effortless Math unique study program provides you with an in-depth focus on the math portion of the exam, helping you master the math skills that students find the most troublesome.
This book contains most common sample questions that are most likely to appear in the mathematics section of the SSAT Upper Level.

Inside the pages of this comprehensive SSAT Upper Level Math book, students can learn basic math operations in a structured manner with a complete study program to help them understand essential math skills. It also has many exciting features, including:

- Dynamic design and easy-to-follow activities
- A fun, interactive and concrete learning process
- Targeted, skill-building practices
- Fun exercises that build confidence
- Math topics are grouped by category, so you can focus on the topics you struggle on
- All solutions for the exercises are included, so you will always find the answers
- 2 Complete SSAT UPPER LEVEL Math Practice Tests that reflect the format and question types on SSAT Upper Level

SSAT Upper Level Mathematics Prep 2019 is an incredibly useful tool for those who want to review all topics being covered on the SSAT Upper Level test. It efficiently and effectively reinforces learning outcomes through engaging questions and repeated practice, helping you to quickly master basic Math skills.

About the Author

Reza Nazari is the author of more than 100 Math learning books including:
– **Math and Critical Thinking Challenges:** For the Middle and High School Student
– **ACT Math in 30 Days.**
– **ASVAB Math Workbook 2018 – 2019**
– **Effortless Math Education Workbooks**
– **and many more Mathematics books ...**

Reza is also an experienced Math instructor and a test–prep expert who has been tutoring students since 2008. Reza is the founder of Effortless Math Education, a tutoring company that has helped many students raise their standardized test scores—and attend the colleges of their dreams. Reza provides an individualized custom learning plan and the personalized attention that makes a difference in how students view math.

You can contact Reza via email at:
reza@EffortlessMath.com

Find Reza's professional profile at:
goo.gl/zoC9rJ

Contents

Chapter 1: Whole Numbers 12
Rounding 13
Whole Number Addition and Subtraction 14
Whole Number Multiplication and Division 15
Rounding and Estimates 16
Answers of Worksheets – Chapter 1 17
Chapter 2: Fractions and Decimals 19
Simplifying Fractions 20
Adding and Subtracting Fractions 21
Multiplying and Dividing Fractions 22
Adding Mixed Numbers 23
Subtract Mixed Numbers 24
Multiplying Mixed Numbers 25
Dividing Mixed Numbers 26
Comparing Decimals 27
Rounding Decimals 28
Adding and Subtracting Decimals 29
Multiplying and Dividing Decimals 30
Converting Between Fractions, Decimals and Mixed Numbers 31
Factoring Numbers 32
Greatest Common Factor 33
Least Common Multiple 34
Divisibility Rules 35
Answers of Worksheets – Chapter 2 36
Chapter 3: Real Numbers and Integers 42
Adding and Subtracting Integers 43
Multiplying and Dividing Integers 44
Ordering Integers and Numbers 45
Arrange, Order, and Comparing Integers 46
Order of Operations 47
Mixed Integer Computations 48

Integers and Absolute Value ... 49
Answers of Worksheets – CHAPTER 3 ... 50
Chapter 4: Proportions and Ratios ... 53
Writing Ratios ... 54
Simplifying Ratios ... 55
Create a Proportion ... 56
Similar Figures ... 57
Simple Interest ... 58
Ratio and Rates Word Problems ... 59
Answers of Worksheets – Chapter 4 ... 60
Chapter 5: Percent ... 62
Percentage Calculations ... 63
Converting Between Percent, Fractions, and Decimals ... 64
Percent Problems ... 65
Markup, Discount, and Tax ... 66
Answers of Worksheets – Chapter 5 ... 67
Chapter 6: Algebraic Expressions ... 69
Expressions and Variables ... 70
Simplifying Variable Expressions ... 71
Simplifying Polynomial Expressions ... 72
Translate Phrases into an Algebraic Statement ... 73
The Distributive Property ... 74
Evaluating One Variable ... 75
Evaluating Two Variables ... 76
Combining like Terms ... 77
Answers of Worksheets – Chapter 6 ... 78
Chapter 7: Equations ... 80
One–Step Equations ... 81
Two–Step Equations ... 82
Multi–Step Equations ... 83
Answers of Worksheets – Chapter 7 ... 84
Chapter 8: Inequalities ... 85

Graphing Single–Variable Inequalities 86
One–Step Inequalities 87
Two–Step Inequalities 88
Multi–Step Inequalities 89
Answers of Worksheets – Chapter 8 90
Chapter 9: Linear Functions 93
Finding Slope 94
Graphing Lines Using Slope–Intercept Form 95
Graphing Lines Using Standard Form 96
Writing Linear Equations 97
Graphing Linear Inequalities 98
Finding Midpoint 99
Finding Distance of Two Points 100
Answers of Worksheets – Chapter 9 101
Chapter 10: Polynomials 105
Classifying Polynomials 106
Writing Polynomials in Standard Form 107
Simplifying Polynomials 108
Adding and Subtracting Polynomials 109
Multiplying Monomials 110
Multiplying and Dividing Monomials 111
Multiplying a Polynomial and a Monomial 112
Multiplying Binomials 113
Factoring Trinomials 114
Operations with Polynomials 115
Answers of Worksheets – Chapter 10 116
Chapter 11: Systems of Equations 120
Solving Systems of Equations by Substitution 121
Solving Systems of Equations by Elimination 122
Systems of Equations Word Problems 123
Answers of Worksheets – Chapter 11 124
Chapter 12: Exponents and Radicals 125

Multiplication Property of Exponents 126
Division Property of Exponents 127
Powers of Products and Quotients 128
Zero and Negative Exponents 129
Negative Exponents and Negative Bases 130
Writing Scientific Notation 131
Square Roots 132
Answers of Worksheets – Chapter 12 133
Chapter 13: Geometry 136
The Pythagorean Theorem 137
Area of Triangles 138
Perimeter of Polygons 139
Area and Circumference of Circles 140
Area of Squares, Rectangles, and Parallelograms 141
Area of Trapezoids 142
Answers of Worksheets – Chapter 13 143
Chapter 14: Solid Figures 144
Volume of Cubes 145
Volume of Rectangle Prisms 146
Surface Area of Cubes 147
Surface Area of a Rectangle Prism 148
Volume of a Cylinder 149
Surface Area of a Cylinder 150
Answers of Worksheets – Chapter 14 151
Chapter 15: Statistics 152
Mean, Median, Mode, and Range of the Given Data 153
Box and Whisker Plots 154
Bar Graph 155
Stem–And–Leaf Plot 156
The Pie Graph or Circle Graph 157
Scatter Plots 158
Probability Problems 159

Answers of Worksheets – Chapter 15 160
SSAT Upper Level Test Review 163
SSAT Upper Level Math Practice Tests 164
SSAT Upper Level Math Practice Test 1 167
SSAT Upper Level Math Practice Test 2 189
SSAT UPPER LEVEL Math Practice Tests Answers and Explanations 210

Chapter 1: Whole Numbers

Topics that you'll learn in this chapter:

- ✓ Rounding
- ✓ Whole Number Addition and Subtraction
- ✓ Whole Number Multiplication and Division
- ✓ Rounding and Estimates

"If people do not believe that mathematics is simple, it is only because they do not realize how complicated life is." — John von Neumann

Rounding

Helpful Hints	– Rounding is putting a number up or down to the nearest whole number or the nearest hundred, etc.	**Example:** 64 rounded to the nearest ten is 60, because 64 is closer to 60 than to 70.

Round each number to the underlined place value.

1) $\underline{9}72$
2) $2{,}\underline{9}95$
3) $3\underline{6}4$
4) $\underline{8}1$
5) $\underline{5}5$
6) $3\underline{3}4$
7) $1{,}\underline{2}03$
8) $9.\underline{5}7$
9) $7.\underline{4}84$
10) $9.\underline{1}4$
11) $\underline{3}9$
12) $\underline{9}{,}123$
13) $3{,}4\underline{5}2$
14) $\underline{5}69$
15) $1{,}\underline{2}30$
16) $\underline{9}8$
17) $\underline{9}3$
18) $\underline{3}7$
19) $4\underline{9}3$
20) $2{,}\underline{9}23$
21) $\underline{9}{,}845$
22) $5\underline{5}5$
23) $\underline{9}39$
24) $\underline{6}9$

Whole Number Addition and Subtraction

Helpful Hints	1– Line up the numbers. 2– Start with the unit place. (ones place) 3– Regroup if necessary. 4– Add or subtract the tens place. 5– Continue with other digits.	**Example:** 231 + 120 = 351 292 – 90 = 202

Solve.

1) A school had 891 students last year. If all last year students and 338 new students have registered for this year, how many students will there be in total?

2) Alice has just started her first job after graduating from college. Her yearly income is $33,000 per year. Alice's father income is $56,000 per year and her mother's income is $49,000. What is yearly income of Alice and her parent altogether?

3) Tom had $895 dollars in his saving account. He gave $235 dollars to his sister, Lisa. How much money does he have left?

4) Emily has 830 marbles, Daniel has 970 marbles, and Ethan has 230 marbles less than Daniel. How many marbles do they have in all?

Find the missing number.

5) 890 – …………… = 300

6) 1000 – …………… = 200

7) …………… – 4000 = 92000

8) 60000 – 51000 = ……………

9) 3400 – …………… = 3200

10) 33000 – 5000 = ……………

Whole Number Multiplication and Division

Helpful Hints		
	Multiplication: – Learn the times tables first! – For multiplication, line up the numbers you are multiplying. – Start with the ones place. – Continue with other digits – A typical division problem: Dividend ÷ Divisor = Quotient **Division:** – In division, we want to find how many times a number (divisor) is contained in another number (dividend). – The result in a division problem is the quotient.	**Example:** $200 \times 90 = 18,000$ $18,000 \div 90 = 200$

Multiply and divided.

1) $340 \div 8 =$
2) $1800 \div 20 =$
3) $50000 \div 10 =$
4) $966 \div 30 =$
5) $201 \times 20 =$
6) $400 \times 50 =$
7) $400 \times 90 =$
8) $888 \times 90 =$
9) $80 \times 80 =$
10) $122 \times 12 =$
11) $609 \times 8 =$
12) $220 \times 12 =$

13) A group of 235 students has collected $8,565 for charity during last month. They decided to split the money evenly among 5 charities. How much will each charity receive?

14) Maria and her two brothers have 9 boxes of crayons. Each box contains 56 crayons. How many crayons do Maria and her two brothers have?

Rounding and Estimates

Helpful Hints

– Rounding and estimating are math strategies used for approximating a number.

– To estimate means to make a rough guess or calculation.

– To round means to simplify a known number by scaling it slightly up or down.

Example:

$73 + 69 \approx 70 + 70 = 140$

Estimate the sum by rounding each added to the nearest ten.

1) 55 + 9

2) 25 + 12

3) 83 + 7

4) 32 + 37

5) 13 + 74

6) 34 + 11

7) 39 + 77

8) 25 + 4

9) 61 + 73

10) 64 + 59

11) 14 + 68

12) 82 + 12

13) 43 + 66

14) 45 + 65

15) 553 + 232

16) 418 + 846

17) 582 + 277

18) 2771 + 1651

19) 7436 + 3575

20) 1542 + 8738

21) 3843 + 6579

22) 4722 + 8186

23) 2419 + 7224

24) 6768 + 3169

Answers of Worksheets – Chapter 1

Rounding

1) 1000
2) 3000
3) 360
4) 80
5) 60
6) 330
7) 1200
8) 9.6
9) 7.5
10) 9.1
11) 40
12) 9000
13) 3,450
14) 600
15) 1,200
16) 100
17) 90
18) 40
19) 490
20) 2,900
21) 10,000
22) 560
23) 900
24) 70

Whole Number Addition and Subtraction

1) 1229
2) 138000
3) 660
4) 2540
5) 590
6) 800
7) 96000
8) 9000
9) 200
10) 28000

Whole Number Multiplication and Division

1) 42.5
2) 90
3) 5000
4) 32.2
5) 4020
6) 20000
7) 36000
8) 79920
9) 6400
10) 1464
11) 4872
12) 2640
13) 1713
14) 504

Rounding and Estimates

1) 70
2) 40
3) 90
4) 70
5) 37
6) 40
7) 120
8) 30
9) 130
10) 120
11) 80
12) 90
13) 110
14) 120
15) 780
16) 1270
17) 860
18) 4420
19) 11020
20) 10280
21) 10420
22) 12910
23) 9640
24) 9940

Chapter 2: Fractions and Decimals

Topics that you'll learn in this chapter:

- ✓ Simplifying Fractions
- ✓ Adding and Subtracting Fractions
- ✓ Multiplying and Dividing Fractions
- ✓ Adding Mixed Numbers
- ✓ Subtract Mixed Numbers
- ✓ Multiplying Mixed Numbers
- ✓ Dividing Mixed Numbers
- ✓ Comparing Decimals
- ✓ Rounding Decimals
- ✓ Adding and Subtracting Decimals
- ✓ Multiplying and Dividing Decimals
- ✓ Converting Between Fractions, Decimals and Mixed Numbers
- ✓ Factoring Numbers
- ✓ Greatest Common Factor
- ✓ Least Common Multiple
- ✓ Divisibility Rules

"A Man is like a fraction whose numerator is what he is and whose denominator is what he thinks of himself. The larger the denominator, the smaller the fraction." –Tolstoy

Simplifying Fractions

Helpful Hints

– Evenly divide both the top and bottom of the fraction by 2, 3, 5, 7, ... etc.
– Continue until you can't go any further.

Example:

$$\frac{4}{12} = \frac{2}{6} = \frac{1}{3}$$

Simplify the fractions.

1) $\frac{22}{36}$

2) $\frac{8}{10}$

3) $\frac{12}{18}$

4) $\frac{6}{8}$

5) $\frac{13}{39}$

6) $\frac{5}{20}$

7) $\frac{16}{36}$

8) $\frac{18}{36}$

9) $\frac{20}{50}$

10) $\frac{6}{54}$

11) $\frac{45}{81}$

12) $\frac{21}{28}$

13) $\frac{35}{56}$

14) $\frac{52}{64}$

15) $\frac{13}{65}$

16) $\frac{44}{77}$

17) $\frac{21}{42}$

18) $\frac{15}{36}$

19) $\frac{9}{24}$

20) $\frac{20}{80}$

21) $\frac{25}{45}$

Adding and Subtracting Fractions

Helpful Hints

– For "like" fractions (fractions with the same denominator), add or subtract the numerators and write the answer over the common denominator.
– Find equivalent fractions with the same denominator before you can add or subtract fractions with different denominators.
– Adding and Subtracting with the same denominator:

$$\frac{a}{b}+\frac{c}{b}=\frac{a+c}{b}$$
$$\frac{a}{b}-\frac{c}{b}=\frac{a-c}{b}$$

– Adding and Subtracting fractions with different denominators:

$$\frac{a}{b}+\frac{c}{d}=\frac{ad+cb}{bd}$$
$$\frac{a}{b}-\frac{c}{d}=\frac{ad-cb}{bd}$$

Add fractions.

1) $\frac{2}{3}+\frac{1}{2}$

2) $\frac{3}{5}+\frac{1}{3}$

3) $\frac{5}{6}+\frac{1}{2}$

4) $\frac{7}{4}+\frac{5}{9}$

5) $\frac{2}{5}+\frac{1}{5}$

6) $\frac{3}{7}+\frac{1}{2}$

7) $\frac{3}{4}+\frac{2}{5}$

8) $\frac{2}{3}+\frac{1}{5}$

9) $\frac{16}{25}+\frac{3}{5}$

Subtract fractions.

10) $\frac{4}{5}-\frac{2}{5}$

11) $\frac{3}{5}-\frac{2}{7}$

12) $\frac{1}{2}-\frac{1}{3}$

13) $\frac{8}{9}-\frac{3}{5}$

14) $\frac{3}{7}-\frac{3}{14}$

15) $\frac{4}{15}-\frac{1}{10}$

16) $\frac{3}{4}-\frac{13}{18}$

17) $\frac{5}{8}-\frac{2}{5}$

18) $\frac{1}{2}-\frac{1}{9}$

Multiplying and Dividing Fractions

Helpful Hints

– **Multiplying fractions:** multiply the top numbers and multiply the bottom numbers.

– **Dividing fractions:** Keep, Change, Flip

Keep first fraction, change division sign to multiplication, and flip the numerator and denominator of the second fraction. Then, solve!

Example:

$\frac{a}{b} \times \frac{c}{d} = \frac{a \times c}{b \times d}$

$\frac{a}{b} \div \frac{c}{d} = \frac{a}{b} \times \frac{d}{c} = \frac{ad}{bc}$

Multiplying fractions. Then simplify.

1) $\frac{1}{5} \times \frac{2}{3}$

2) $\frac{3}{4} \times \frac{2}{3}$

3) $\frac{2}{5} \times \frac{3}{7}$

4) $\frac{3}{8} \times \frac{1}{3}$

5) $\frac{3}{5} \times \frac{2}{5}$

6) $\frac{7}{9} \times \frac{1}{3}$

7) $\frac{2}{3} \times \frac{3}{8}$

8) $\frac{1}{4} \times \frac{1}{3}$

9) $\frac{5}{7} \times \frac{7}{12}$

Dividing fractions.

10) $\frac{2}{9} \div \frac{1}{4}$

11) $\frac{1}{2} \div \frac{1}{3}$

12) $\frac{6}{11} \div \frac{3}{4}$

13) $\frac{11}{14} \div \frac{1}{10}$

14) $\frac{3}{5} \div \frac{5}{9}$

15) $\frac{1}{2} \div \frac{1}{2}$

16) $\frac{3}{5} \div \frac{1}{5}$

17) $\frac{12}{21} \div \frac{3}{7}$

18) $\frac{5}{14} \div \frac{9}{10}$

Adding Mixed Numbers

Helpful Hints

Use the following steps for both adding and subtracting mixed numbers.

- Find the Least Common Denominator (LCD)
- Find the equivalent fractions for each mixed number.
- Add fractions after finding common denominator.
- Write your answer in lowest terms.

Example:

$1\frac{3}{4} + 2\frac{3}{8} = 4\frac{1}{8}$

Add.

1) $4\frac{1}{2} + 5\frac{1}{2}$

2) $2\frac{3}{8} + 3\frac{1}{8}$

3) $6\frac{1}{5} + 3\frac{2}{5}$

4) $1\frac{1}{3} + 2\frac{2}{3}$

5) $5\frac{1}{6} + 5\frac{1}{2}$

6) $3\frac{1}{3} + 1\frac{1}{3}$

7) $1\frac{10}{11} + 1\frac{1}{3}$

8) $2\frac{3}{6} + 1\frac{1}{2}$

9) $5\frac{3}{5} + 5\frac{1}{5}$

10) $7 + \frac{1}{5}$

11) $1\frac{5}{7} + \frac{1}{3}$

12) $2\frac{1}{4} + 1\frac{1}{2}$

Subtract Mixed Numbers

Helpful Hints

Use the following steps for both adding and subtracting mixed numbers.

Find the Least Common Denominator (LCD)
- Find the equivalent fractions for each mixed number.
- Add or subtract fractions after finding common denominator.
- Write your answer in lowest terms.

Example:

$5\frac{2}{3} - 3\frac{2}{7} = 2\frac{8}{21}$

Subtract.

1) $4\frac{1}{2} - 3\frac{1}{2}$

2) $3\frac{3}{8} - 3\frac{1}{8}$

3) $6\frac{3}{5} - 5\frac{1}{5}$

4) $2\frac{1}{3} - 1\frac{2}{3}$

5) $6\frac{1}{6} - 5\frac{1}{2}$

6) $3\frac{1}{3} - 1\frac{1}{3}$

7) $2\frac{10}{11} - 1\frac{1}{3}$

8) $2\frac{1}{2} - 1\frac{1}{2}$

9) $6\frac{3}{5} - 2\frac{1}{5}$

10) $7\frac{2}{5} - 1\frac{1}{5}$

11) $2\frac{5}{7} - 1\frac{1}{3}$

12) $2\frac{1}{4} - 1\frac{1}{2}$

Multiplying Mixed Numbers

Helpful Hints

1- Convert the mixed numbers to improper fractions.
2- Multiply fractions and simplify if necessary.

$$a\frac{c}{b} = a + \frac{c}{b} = \frac{ab + c}{b}$$

Example:

$2\frac{1}{3} \times 5\frac{3}{7} =$

$\frac{7}{3} \times \frac{38}{7} = \frac{38}{3} = 12\frac{2}{3}$

Find each product.

1) $1\frac{2}{3} \times 1\frac{1}{4}$

2) $1\frac{3}{5} \times 1\frac{2}{3}$

3) $1\frac{2}{3} \times 3\frac{2}{7}$

4) $4\frac{1}{8} \times 1\frac{2}{5}$

5) $2\frac{2}{5} \times 3\frac{1}{5}$

6) $1\frac{1}{3} \times 1\frac{2}{3}$

7) $1\frac{5}{8} \times 2\frac{1}{2}$

8) $3\frac{2}{5} \times 2\frac{1}{5}$

9) $2\frac{2}{3} \times 4\frac{1}{4}$

10) $2\frac{3}{5} \times 1\frac{2}{4}$

11) $1\frac{1}{3} \times 1\frac{1}{4}$

12) $3\frac{2}{5} \times 1\frac{1}{5}$

Dividing Mixed Numbers

Helpful Hints

1- Convert the mixed numbers to improper fractions.
2- Divide fractions and simplify if necessary.

$a\frac{c}{b} = a + \frac{c}{b} = \frac{ab+c}{b}$

Example:

$10\frac{1}{2} \div 5\frac{3}{5} =$

$\frac{21}{2} \div \frac{28}{5} = \frac{21}{2} \times \frac{5}{28} = \frac{105}{56}$

$= 1\frac{7}{8}$

Find each quotient.

1) $2\frac{1}{5} \div 2\frac{1}{2}$

2) $2\frac{3}{5} \div 1\frac{1}{3}$

3) $3\frac{1}{6} \div 4\frac{2}{3}$

4) $1\frac{2}{3} \div 3\frac{1}{3}$

5) $4\frac{1}{8} \div 2\frac{2}{4}$

6) $3\frac{1}{2} \div 2\frac{3}{5}$

7) $3\frac{5}{9} \div 1\frac{2}{5}$

8) $2\frac{2}{7} \div 1\frac{1}{2}$

9) $3\frac{1}{5} \div 1\frac{1}{2}$

10) $4\frac{3}{5} \div 2\frac{1}{3}$

11) $6\frac{1}{6} \div 1\frac{2}{3}$

12) $2\frac{2}{3} \div 1\frac{1}{3}$

Comparing Decimals

Helpful Hints

- **Decimals:** is a fraction written in a special form. For example, instead of writing $\frac{1}{2}$ you can write 0.5.
- **For comparing:**

  Equal to =

  Less than <

  Greater than >

  Greater than or equal ≥

  Less than or equal ≤

Example:

2.67 > 0.267

Write the correct comparison symbol (>, < or =).

1) 1.25 2.3

2) 0.5 0.23

3) 3.2 3.2

4) 4.58 45.8

5) 2.75 0.275

6) 5.2 5

7) 3.1 0.31

8) 6.33 0.733

9) 8 0.8

10) 4.56 0.456

11) 1.12 1.14

12) 2.77 2.78

13) 6.08 6.11

14) 1.11 0.211

15) 2.6 2.55

16) 1.24 1.25

17) 5.52 0.552

18) 0.33 0.033

19) 14.4 14.4

20) 0.05 0.50

21) 0.59 0.7

22) 0.5 0.05

23) 0.90 0.9

24) 0.27 0.4

Rounding Decimals

Helpful Hints

We can round decimals to a certain accuracy or number of decimal places. This is used to make calculation easier to do and results easier to understand, when exact values are not too important.

First, you'll need to remember your place values:

Example:

$\underline{6}.37 = 6$

12.4567

1: tens   2: ones   4: tenths

5: hundredths   6: thousandths   7: tens thousandths

Round each decimal number to the nearest place indicated.

1) $0.\underline{2}3$

2) $4.\underline{0}4$

3) $5.\underline{6}23$

4) $0.\underline{2}66$

5) $\underline{6}.37$

6) $0.\underline{8}8$

7) $8.\underline{2}4$

8) $\underline{7}.0760$

9) $1.6\underline{2}9$

10) $6.\underline{3}959$

11) $\underline{1}.9$

12) $\underline{5}.2167$

13) $5.\underline{8}63$

14) $8.\underline{5}4$

15) $8\underline{0}.69$

16) $6\underline{5}.85$

17) $70.\underline{7}8$

18) $61\underline{5}.755$

19) $1\underline{6}.4$

20) $9\underline{5}.81$

21) $\underline{2}.408$

22) $7\underline{6}.3$

23) $116.\underline{5}14$

24) $8.\underline{0}6$

Adding and Subtracting Decimals

Helpful Hints

1– Line up the numbers.

2– Add zeros to have same number of digits for both numbers.

3– Add or Subtract using column addition or subtraction.

Example:

$$\begin{array}{r} 16.18 \\ -\ 13.45 \\ \hline 2.73 \end{array}$$

Add and subtract decimals.

1) $\begin{array}{r} 15.14 \\ -\ 12.18 \\ \hline \end{array}$ ______

3) $\begin{array}{r} 82.56 \\ +\ 12.28 \\ \hline \end{array}$ ______

5) $\begin{array}{r} 90.37 \\ +\ 56.97 \\ \hline \end{array}$ ______

2) $\begin{array}{r} 65.72 \\ +\ 43.67 \\ \hline \end{array}$ ______

4) $\begin{array}{r} 34.18 \\ -\ 23.45 \\ \hline \end{array}$ ______

6) $\begin{array}{r} 45.78 \\ -\ 23.39 \\ \hline \end{array}$ ______

Solve.

7) _____ + 1.3 = 4.8

8) 4.2 + _____ = 11.6

9) 9.9 + _____ = 16

10) 6.9 + _____ = 16.4

11) _____ + 5.1 = 8.6

12) _____ + 7.9 = 15.2

Multiplying and Dividing Decimals

Helpful Hints

For Multiplication:

– Set up and multiply the numbers as you do with whole numbers.

– Count the total number of decimal places in both of the factors.

– Place the decimal point in the product.

For Division:

– If the divisor is not a whole number, move decimal point to right to make it a whole number. Do the same for dividend.

– Divide similar to whole numbers.

Find each product.

1) $\begin{array}{r} 4.5 \\ \times\ 1.6 \\ \hline \end{array}$

2) $\begin{array}{r} 7.7 \\ \times\ 9.9 \\ \hline \end{array}$

3) $\begin{array}{r} 2.6 \\ \times\ 1.5 \\ \hline \end{array}$

4) $\begin{array}{r} 8.9 \\ \times\ 9.7 \\ \hline \end{array}$

5) $\begin{array}{r} 15.1 \\ \times\ 12.6 \\ \hline \end{array}$

6) $\begin{array}{r} 6.9 \\ \times\ 3.3 \\ \hline \end{array}$

7) $\begin{array}{r} 5.7 \\ \times\ 7.8 \\ \hline \end{array}$

8) $\begin{array}{r} 98.20 \\ \times\ 100 \\ \hline \end{array}$

9) $\begin{array}{r} 23.99 \\ \times\ 1000 \\ \hline \end{array}$

Find each quotient.

10) 9.2 ÷ 3.6

11) 27.6 ÷ 3.8

12) 12.6 ÷ 4.7

13) 6.5 ÷ 8.1

14) 1. 4 ÷ 10

15) 3.6 ÷ 100

16) 4.24 ÷ 10

17) 14.6 ÷ 100

18) 1.8 ÷ 1000

Converting Between Fractions, Decimals and Mixed Numbers

Helpful Hints

Fraction to Decimal:

– Divide the top number by the bottom number.

Decimal to Fraction:

– Write decimal over 1.

– Multiply both top and bottom by 10 for every digit on the right side of the decimal point.

– Simplify.

Convert fractions to decimals.

1) $\frac{9}{10}$

4) $\frac{2}{5}$

7) $\frac{12}{10}$

2) $\frac{56}{100}$

5) $\frac{3}{9}$

8) $\frac{8}{5}$

3) $\frac{3}{4}$

6) $\frac{40}{50}$

9) $\frac{69}{10}$

Convert decimal into fraction or mixed numbers.

10) 0.3

14) 0.8

18) 0.08

11) 4.5

15) 0.25

19) 0.45

12) 2.5

16) 0.14

20) 2.6

13) 2.3

17) 0.2

21) 5.2

Factoring Numbers

Helpful Hints

- Factoring numbers means to break the numbers into their prime factors.
- First few prime numbers: 2, 3, 5, 7, 11, 13, 17, 19

Example:

$12 = 2 \times 2 \times 3$

List all positive factors of each number.

1) 68	6) 78	11) 54
2) 56	7) 50	12) 28
3) 24	8) 98	13) 55
4) 40	9) 45	14) 85
5) 86	10) 26	15) 48

List the prime factorization for each number.

16) 50	19) 21	22) 26
17) 25	20) 45	23) 86
18) 69	21) 68	24) 93

Greatest Common Factor

Helpful Hints

- List the prime factors of each number.
- Multiply common prime factors.

Example:

$200 = 2 \times 2 \times 2 \times 5 \times 5$

$60 = 2 \times 2 \times 3 \times 5$

GCF (200, 60) $= 2 \times 2 \times 5 = 20$

Find the GCF for each number pair.

1) 20, 30
2) 4, 14
3) 5, 45
4) 68, 12
5) 5, 12
6) 15, 27
7) 3, 24
8) 34, 6
9) 4, 10
10) 5, 3
11) 6, 16
12) 30, 3
13) 24, 28
14) 70, 10
15) 45, 8
16) 90, 35
17) 78, 34
18) 55, 75
19) 60, 72
20) 100, 78
21) 30, 40

Least Common Multiple

Helpful Hints	- Find the GCF for the two numbers. - Divide that GCF into either number. - Take that answer and multiply it by the other number.	**Example:** LCM (200, 60): GCF is 20 $200 \div 20 = 10$ $10 \times 60 = 600$

Find the LCM for each number pair.

1) 4, 14
2) 5, 15
3) 16, 10
4) 4, 34
5) 8, 3
6) 12, 24
7) 9, 18
8) 5, 6
9) 8, 19
10) 9, 21
11) 19, 29
12) 7, 6
13) 25, 6
14) 4, 8
15) 30, 10, 50
16) 18, 36, 27
17) 12, 8, 18
18) 8, 18, 4
19) 26, 20, 30
20) 10, 4, 24
21) 15, 30, 45

Divisibility Rules

Helpful Hints

- Divisibility means that a number can be divided by other numbers evenly.

Example:

24 is divisible by 6, because 24 ÷ 6 = 4

Use the divisibility rules to find the factors of each number.

8	2̲ 3 4̲ 5 6 7 8̲ 9 10
1) 16	2 3 4 5 6 7 8 9 10
2) 10	2 3 4 5 6 7 8 9 10
3) 15	2 3 4 5 6 7 8 9 10
4) 28	2 3 4 5 6 7 8 9 10
5) 36	2 3 4 5 6 7 8 9 10
6) 15	2 3 4 5 6 7 8 9 10
7) 27	2 3 4 5 6 7 8 9 10
8) 70	2 3 4 5 6 7 8 9 10
9) 57	2 3 4 5 6 7 8 9 10
10) 102	2 3 4 5 6 7 8 9 10
11) 144	2 3 4 5 6 7 8 9 10
12) 75	2 3 4 5 6 7 8 9 10

Answers of Worksheets – Chapter 2

Simplifying Fractions

1) $\frac{11}{18}$

2) $\frac{4}{5}$

3) $\frac{2}{3}$

4) $\frac{3}{4}$

5) $\frac{1}{3}$

6) $\frac{1}{4}$

7) $\frac{4}{9}$

8) $\frac{1}{2}$

9) $\frac{2}{5}$

10) $\frac{1}{9}$

11) $\frac{5}{9}$

12) $\frac{3}{4}$

13) $\frac{5}{8}$

14) $\frac{13}{16}$

15) $\frac{1}{5}$

16) $\frac{4}{7}$

17) $\frac{1}{2}$

18) $\frac{5}{12}$

19) $\frac{3}{8}$

20) $\frac{1}{4}$

21) $\frac{5}{9}$

Adding and Subtracting Fractions

1) $\frac{7}{6}$

2) $\frac{14}{15}$

3) $\frac{4}{3}$

4) $\frac{83}{36}$

5) $\frac{3}{5}$

6) $\frac{13}{14}$

7) $\frac{23}{20}$

8) $\frac{13}{15}$

9) $\frac{31}{25}$

10) $\frac{2}{5}$

11) $\frac{11}{35}$

12) $\frac{1}{6}$

13) $\frac{13}{45}$

14) $\frac{3}{14}$

15) $\frac{1}{6}$

16) $\frac{1}{36}$

17) $\frac{9}{40}$

18) $\frac{7}{18}$

Multiplying and Dividing Fractions

1) $\frac{2}{15}$
2) $\frac{1}{2}$
3) $\frac{6}{35}$
4) $\frac{1}{8}$
5) $\frac{6}{25}$
6) $\frac{7}{27}$
7) $\frac{1}{4}$
8) $\frac{1}{12}$
9) $\frac{5}{12}$
10) $\frac{8}{9}$
11) $\frac{3}{2}$
12) $\frac{8}{11}$
13) $\frac{55}{7}$
14) $\frac{27}{25}$
15) 1
16) 3
17) $\frac{4}{3}$
18) $\frac{25}{63}$

Adding Mixed Numbers

1) 10
2) $5\frac{1}{2}$
3) $9\frac{3}{5}$
4) 4
5) $10\frac{2}{3}$
6) $4\frac{2}{3}$
7) $3\frac{8}{33}$
8) 4
9) $10\frac{4}{5}$
10) $7\frac{1}{5}$
11) $2\frac{1}{21}$
12) $3\frac{3}{4}$

Subtract Mixed Numbers

1) 1
2) $\frac{1}{4}$
3) $1\frac{2}{5}$
4) $\frac{2}{3}$
5) $\frac{2}{3}$
6) 2
7) $1\frac{19}{33}$
8) 1
9) $4\frac{2}{5}$
10) $6\frac{1}{5}$
11) $1\frac{8}{21}$
12) $\frac{3}{4}$

Multiplying Mixed Numbers

1) $2\frac{1}{12}$
2) $2\frac{2}{3}$
3) $5\frac{10}{21}$
4) $5\frac{31}{40}$
5) $7\frac{17}{25}$
6) $2\frac{2}{9}$
7) $4\frac{1}{16}$
8) $7\frac{12}{25}$
9) $11\frac{1}{3}$
10) $3\frac{9}{10}$
11) $1\frac{2}{3}$
12) $4\frac{2}{25}$

Dividing Mixed Numbers

1) $\frac{22}{25}$
2) $1\frac{19}{20}$
3) $\frac{19}{28}$
4) $\frac{1}{2}$
5) $1\frac{13}{20}$
6) $1\frac{9}{26}$
7) $2\frac{34}{63}$
8) $1\frac{11}{21}$
9) $2\frac{2}{15}$
10) $1\frac{34}{35}$
11) $3\frac{7}{10}$
12) 2

Comparing Decimals

1) 1.25 < 2.3
2) 0.5 > 0.23
3) 3.2 = 3.2
4) 4.58 < 45.8
5) 2.75 > 0.275
6) 5.2 > 5
7) 3.1 > 0.31
8) 6.33 > 0.733
9) 8 > 0.8
10) 4.56 > 0.456
11) 1.12 < 1.14
12) 2.77 < 2.78
13) 6.08 < 6.11
14) 1.11 > 0.211
15) 2.6 > 2.55
16) 1.24 < 1.25
17) 5.52 > 0.552
18) 0.33 > 0.033
19) 14.4 = 14.4
20) 0.05 < 0.50
21) 0.59 < 0.7
22) 0.5 > 0.05
23) 0.90 = 0.9
24) 0.27 < 0.4

Rounding Decimals

1) 0.2
2) 4.0
3) 5.6
4) 0.3
5) 6
6) 0.9
7) 8.2
8) 7
9) 1.63
10) 6.4
11) 2
12) 5
13) 5.9
14) 8.5
15) 81
16) 66
17) 70.8
18) 616
19) 16
20) 96
21) 2
22) 76
23) 116.5
24) 8.1

Adding and Subtracting Decimals

1) 2.96
2) 109.39
3) 94.84
4) 10.73
5) 147.34
6) 22.39
7) 3.5
8) 7.4
9) 6.1
10) 9.5
11) 3.5
12) 7.3

Multiplying and Dividing Decimals

1) 7.2
2) 76.23
3) 3.9
4) 86.33
5) 190.26
6) 22.77
7) 44.46
8) 9820
9) 23990
10) 2.5555…
11) 7.2631…
12) 2.6808…
13) 0.8024…
14) 0.14
15) 0.036
16) 0.424
17) 0.146
18) 0.0018

Converting Between Fractions, Decimals and Mixed Numbers

1) 0.9
2) 0.56
3) 0.75
4) 0.4
5) 0.333…
6) 0.8
7) 1.2
8) 1.6
9) 6.9
10) $\frac{3}{10}$
11) $4\frac{1}{2}$
12) $2\frac{1}{2}$
13) $2\frac{3}{10}$
14) $\frac{4}{5}$
15) $\frac{1}{4}$

16) $\frac{7}{50}$
17) $\frac{1}{5}$
18) $\frac{2}{25}$
19) $\frac{9}{20}$
20) $2\frac{3}{5}$
21) $5\frac{1}{5}$

Factoring Numbers

1) 1, 2, 4, 17, 34, 68
2) 1, 2, 4, 7, 8, 14, 28, 56
3) 1, 2, 3, 4, 6, 8, 12, 24
4) 1, 2, 4, 5, 8, 10, 20, 40
5) 1, 2, 43, 86
6) 1, 2, 3, 6, 13, 26, 39, 78
7) 1, 2, 5, 10, 25, 50
8) 1, 2, 7, 14, 49, 98
9) 1, 3, 5, 9, 15, 45
10) 1, 2, 13, 26
11) 1, 2, 3, 6, 9, 18, 27, 54
12) 1, 2, 4, 7, 14, 28
13) 1, 5, 11, 55
14) 1, 5, 17, 85
15) 1, 2, 3, 4, 6, 8, 12, 16, 24, 48
16) $2 \times 5 \times 5$
17) 5×5
18) 3×23
19) 3×7
20) $3 \times 3 \times 5$
21) $2 \times 2 \times 17$
22) 2×13
23) 2×43
24) 3×31

Greatest Common Factor

1) 10
2) 2
3) 5
4) 4
5) 1
6) 3
7) 3
8) 2
9) 2
10) 1
11) 2
12) 3
13) 4
14) 10
15) 1
16) 5
17) 2
18) 5
19) 12
20) 2
21) 10

Least Common Multiple

1) 28
2) 15
3) 80
4) 68
5) 24
6) 24
7) 18
8) 30
9) 152
10) 63
11) 551
12) 42
13) 150
14) 8
15) 150
16) 108
17) 72
18) 72
19) 780
20) 120
21) 90

Divisibility Rules

1) 16	$\underline{2}$ 3 $\underline{4}$ 5 6 7 $\underline{8}$ 9 10
2) 10	$\underline{2}$ 3 4 $\underline{5}$ 6 7 8 9 $\underline{10}$
3) 15	2 $\underline{3}$ 4 $\underline{5}$ 6 7 8 9 10
4) 28	$\underline{2}$ 3 $\underline{4}$ 5 6 $\underline{7}$ 8 9 10
5) 36	$\underline{2}$ $\underline{3}$ $\underline{4}$ 5 $\underline{6}$ 7 8 $\underline{9}$ 10
6) 18	$\underline{2}$ $\underline{3}$ 4 5 $\underline{6}$ 7 8 $\underline{9}$ 10
7) 27	2 $\underline{3}$ 4 5 6 7 8 $\underline{9}$ 10
8) 70	$\underline{2}$ 3 4 $\underline{5}$ 6 $\underline{7}$ 8 9 $\underline{10}$
9) 57	2 $\underline{3}$ 4 5 6 7 8 9 10
10) 102	$\underline{2}$ $\underline{3}$ 4 5 $\underline{6}$ 7 8 9 10
11) 144	$\underline{2}$ $\underline{3}$ $\underline{4}$ 5 $\underline{6}$ 7 $\underline{8}$ $\underline{9}$ 10
12) 75	2 $\underline{3}$ 4 $\underline{5}$ 6 7 8 9 10

Chapter 3: Real Numbers and Integers

Topics that you'll learn in this chapter:

- ✓ Adding and Subtracting Integers
- ✓ Multiplying and Dividing Integers
- ✓ Ordering Integers and Numbers
- ✓ Arrange and Order, Comparing Integers
- ✓ Order of Operations
- ✓ Mixed Integer Computations
- ✓ Integers and Absolute Value

"Wherever there is number, there is beauty." –Proclus

Adding and Subtracting Integers

Helpful Hints	- **Integers:** {… , −3, −2, −1, 0, 1, 2, 3, …} Includes: zero, counting numbers, and the negative of the counting numbers. – Add a positive integer by moving to the right on the number line. – Add a negative integer by moving to the left on the number line. – Subtract an integer by adding its opposite.	**Example:** 12 + 10 = 22 25 – 13 = 12 (−24) + 12 = −12 (−14) + (−12) = −26 14 – (−13) = 27

Find the sum.

1) (– 12) + (– 4)
2) 5 + (– 24)
3) (– 14) + 23
4) (– 8) + (39)
5) 43 + (–12)
6) (– 23) + (– 4) + 3
7) 4 + (– 12) + (– 10) + (– 25)
8) 19 + (– 15) + 25 + 11
9) (– 9) + (– 12) + (32 – 14)
10) 4 + (– 30) + (45 – 34)

Find the difference.

11) (– 14) – (– 9) – (18)
12) (– 9) – (– 25)
13) (– 12) – (8)
14) (28) – (– 4)
15) (34) – (2)
16) (55) – (– 5) + (– 4)
17) (9) – (2) – (– 5)
18) (2) – (4) – (– 15)
19) (23) – (4) – (– 34)
20) (– 45) – (– 87)

Multiplying and Dividing Integers

Helpful Hints

(negative) × (negative) = positive

(negative) ÷ (negative) = positive

(negative) × (positive) = negative

(negative) ÷ (positive) = negative

(positive) × (positive) = positive

Examples:

$3 \times 2 = 6$

$3 \times -3 = -9$

$-2 \times -2 = 4$

$10 \div 2 = 5$

$-4 \div 2 = -2$

$-12 \div -6 = 3$

Find each product.

1) $(-8) \times (-2)$
2) 3×6
3) $(-4) \times 5 \times (-6)$
4) $2 \times (-6) \times (-6)$
5) $11 \times (-12)$
6) $10 \times (-5)$
7) 8×8
8) $(-8) \times (-9)$
9) $6 \times (-5) \times 3$
10) $6 \times (-1) \times 2$

Find each quotient.

11) $18 \div 3$
12) $(-24) \div 4$
13) $(-63) \div (-9)$
14) $54 \div 9$
15) $20 \div (-2)$
16) $(-66) \div (-11)$
17) $64 \div 8$
18) $(-121) \div 11$
19) $72 \div 9$
20) $16 \div 4$

Ordering Integers and Numbers

Helpful Hints

To compare numbers, you can use number line! As you move from left to right on the number line, you find a bigger number!

Example:

Order integers from least to greatest.

$(-11, -13, 7, -2, 12)$

$-13 < -11 < -2 < 7 < 12$

Order each set of integers from least to greatest.

1) $-15, -19, 20, -4, 1$ ___, ___, ___, ___, ___, ___

2) $6, -5, 4, -3, 2$ ___, ___, ___, ___, ___, ___

3) $15, -42, 19, 0, -22$ ___, ___, ___, ___, ___, ___

4) $26, -91, 0, -13, 67, -55$ ___, ___, ___, ___, ___, ___

5) $-17, -71, 90, -25, -54, -39$ ___, ___, ___, ___, ___, ___

6) $98, 5, 46, 19, 77, 24$ ___, ___, ___, ___, ___, ___

Order each set of integers from greatest to least.

7) $-2, 5, -3, 6, -4$ ___, ___, ___, ___, ___, ___

8) $-37, 7, -17, 27, 47$ ___, ___, ___, ___, ___, ___

9) $32, -27, 19, -17, 15$ ___, ___, ___, ___, ___, ___

10) $68, 81, 21, -18, 94, 72$ ___, ___, ___, ___, ___, ___

Arrange, Order, and Comparing Integers

Helpful Hints	When using a number line, numbers increase as you move to the right.	**Examples:** $5 < 7$, $-5 < -2$ $-18 < -12$

Arrange these integers in descending order.

1) 21, 71, – 18, – 10, 82 ___, ___, ___, ___, ___, ___

2) 15, 11, 20, 12, – 9, – 5 ___, ___, ___, ___, ___, ___

3) – 5, 20, 15, 9, –11 ___, ___, ___, ___, ___, ___

4) 19, 18, – 9, – 6, – 11 ___, ___, ___, ___, ___, ___

5) 56, – 34, – 12, – 5, 32 ___, ___, ___, ___, ___, ___

Compare. Use >, =, <

6) – 8 ____ 12

7) – 10 ____ –16

8) 43 ____ 34

9) 15 ____ –16

10) – 354 ____ –345

11) – 56 ____ – 58

12) 78 ____ 87

13) – 92 ____ – 102

14) – 12 ____ – 12

15) – 721 ____ – 821

Order of Operations

Helpful Hints

- Use "order of operations" rule when there are more than one math operation.
- PEMDAS (parentheses / exponents / multiply / divide / add / subtract)

Example:

$(12 + 4) \div (-4) = -4$

Evaluate each expression.

1) $(2 \times 2) + 5$

2) $24 - (3 \times 3)$

3) $(6 \times 4) + 8$

4) $25 - (4 \times 2)$

5) $(6 \times 5) + 3$

6) $64 - (2 \times 4)$

7) $25 + (1 \times 8)$

8) $(6 \times 7) + 7$

9) $48 \div (4 + 4)$

10) $(7 + 11) \div (-2)$

11) $9 + (2 \times 5) + 10$

12) $(5 + 8) \times \frac{3}{5} + 2$

13) $2 \times 7 - (\frac{10}{9 - 4})$

14) $(12 + 2 - 5) \times 7 - 1$

15) $(\frac{7}{5 - 1}) \times (2 + 6) \times 2$

16) $20 \div (4 - (10 - 8))$

17) $\frac{50}{4\,(5 - 4) - 3}$

18) $2 + (8 \times 2)$

Mixed Integer Computations

Helpful Hints	It worth remembering:	Example:
	(negative) × (negative) = positive	
	(negative) ÷ (negative) = positive	$(-5) + 6 = 1$
	(negative) × (positive) = negative	$(-3) \times (-2) = 6$
	(negative) ÷ (positive) = negative	$(9) \div (-3) = -3$
	(positive) × (positive) = positive	

Compute.

1) $(-70) \div (-5)$

2) $(-14) \times 3$

3) $(-4) \times (-15)$

4) $(-65) \div 5$

5) $18 \times (-7)$

6) $(-12) \times (-2)$

7) $\frac{(-60)}{(-20)}$

8) $24 \div (-8)$

9) $22 \div (-11)$

10) $\frac{(-27)}{3}$

11) $4 \times (-4)$

12) $\frac{(-48)}{12}$

13) $(-14) \times (-2)$

14) $(-7) \times (7)$

15) $\frac{-30}{-6}$

16) $(-54) \div 6$

17) $(-60) \div (-5)$

18) $(-7) \times (-12)$

19) $(-14) \times 5$

20) $88 \div (-8)$

Integers and Absolute Value

Helpful Hints

To find an absolute value of a number, just find it's distance from 0!

Example:

$|-6| = 6$

$|6| = 6$

$|-12| = 12$

$|12| = 12$

Write absolute value of each number.

1) -4

2) -7

3) -8

4) 4

5) 5

6) -10

7) 1

8) 6

9) 8

10) -2

11) -1

12) 10

13) 3

14) 7

15) -5

16) -3

17) -9

18) 2

19) 4

20) -6

21) 9

Evaluate.

22) $|-43| - |12| + 10$

23) $76 + |-15 - 45| - |3|$

24) $30 + |-62| - 46$

25) $|32| - |-78| + 90$

26) $|-35 + 4| + 6 - 4$

27) $|-4| + |-11|$

28) $|-6 + 3 - 4| + |7 + 7|$

29) $|-9| + |-19| - 5$

Answers of Worksheets – CHAPTER 3

Adding and Subtracting Integers

1) – 16
2) – 19
3) 9
4) 31
5) 31
6) – 24
7) – 43
8) 40
9) – 3
10) – 15
11) – 23
12) 16
13) – 20
14) 32
15) 32
16) 56
17) 12
18) 13
19) 53
20) 42

Multiplying and Dividing Integers

1) 16
2) 18
3) 120
4) 72
5) – 132
6) – 50
7) 64
8) 72
9) – 90
10) – 12
11) 6
12) – 6
13) 7
14) 6
15) – 10
16) 6
17) 8
18) – 11
19) 8
20) 4

Ordering Integers and Numbers

1) – 19, – 15, – 4, 1, 20
2) – 5, – 3, 2, 4, 6
3) – 42, – 22, 0, 15, 19
4) – 91, – 55, – 13, 0, 26, 67
5) – 71, – 54, – 39, – 25, – 17, 90
6) 5, 19, 24, 46, 77, 98
7) 6, 5, – 2, – 3, – 4
8) 47, 27, 7, – 17, – 37
9) 32, 19, 15, – 17, – 27
10) 94, 81, 72, 68, 21, – 18

Arrange and Order, Comparing Integers

1) 82, 71, 21, – 10, – 18
2) 20, 15, 12, 11, – 5, – 9
3) 20, 15, 9, – 5, –11
4) 19, 18, – 6, – 9, – 11
5) 56, 32, – 5, – 12, – 34
6) <
7) >
8) >
9) >
10) <
11) >
12) <
13) >
14) =
15) >

Order of Operations

1) 9
2) 15
3) 32
4) 17
5) 33
6) 56
7) 33
8) 49
9) 6
10) – 9
11) 29
12) 9.8
13) 12
14) 62
15) 28
16) 10
17) 50
18) 18

Mixed Integer Computations

1) 14
2) – 42
3) 60
4) – 13
5) – 126
6) 24
7) 3
8) – 3
9) – 2
10) – 9
11) – 16
12) – 4
13) 28
14) – 49
15) 5
16) – 9
17) 12
18) 84
19) – 70
20) – 11

Integers and Absolute Value

1) 4
2) 7
3) 8
4) 4
5) 5
6) 10
7) 1
8) 6
9) 8
10) 2
11) 1
12) 10
13) 3
14) 7
15) 5
16) 3
17) 9
18) 2
19) 4
20) 6
21) 9
22) 41
23) 133
24) 46
25) 44
26) 33
27) 15
28) 21
29) 23

Chapter 4: Proportions and Ratios

Topics that you'll learn in this chapter:

- ✓ Writing Ratios
- ✓ Simplifying Ratios
- ✓ Create a Proportion
- ✓ Similar Figures
- ✓ Simple Interest
- ✓ Ratio and Rates Word Problems

"Do not worry about your difficulties in mathematics. I can assure you mine are still greater."

– Albert Einstein

Writing Ratios

Helpful Hints	– A ratio is a comparison of two numbers. Ratio can be written as a division.	**Example:** $3:5$, or $\frac{3}{5}$

Express each ratio as a rate and unite rate.

1) 120 miles on 4 gallons of gas.

2) 24 dollars for 6 books.

3) 200 miles on 14 gallons of gas

4) 24 inches of snow in 8 hours

Express each ratio as a fraction in the simplest form.

5) 3 feet out of 30 feet

6) 18 cakes out of 42 cakes

7) 16 dimes t0 24 dimes

8) 12 dimes out of 48 coins

9) 14 cups to 84 cups

10) 45 gallons to 65 gallons

11) 10 miles out of 40 miles

12) 22 blue cars out of 55 cars

13) 32 pennies to 300 pennies

14) 24 beetles out of 86 insects

Simplifying Ratios

Helpful Hints	– You can calculate equivalent ratios by multiplying or dividing both sides of the ratio by the same number.	**Examples:** 3 : 6 = 1 : 2 4 : 9 = 8 : 18

Reduce each ratio.

1) 21 : 49
2) 20 : 40
3) 10: 50
4) 14: 18
5) 45: 27
6) 49: 21
7) 100: 10
8) 12 : 8
9) 35 : 45
10) 8: 20
11) 25: 35
12) 21 : 27
13) 52 : 82
14) 12: 36
15) 24 : 3
16) 15: 30
17) 3 : 36
18) 8 : 16
19) 6 : 100
20) 2 : 20
21) 10: 60
22) 14: 63
23) 68: 80
24) 8: 80

Create a Proportion

Helpful Hints

– A proportion contains 2 equal fractions! A proportion simply means that two fractions are equal.

Example:

2, 4, 8, 16

$\frac{2}{4} = \frac{8}{16}$

Create proportion from the given set of numbers.

1) 1, 6, 2, 3

2) 12, 144, 1, 12

3) 16, 4, 8, 2

4) 9, 5, 27, 15

5) 7, 10, 60, 42

6) 8, 7, 24, 21

7) 10, 5, 8, 4

8) 3, 12, 8, 2

9) 2, 2, 1, 4

10) 3, 6, 7, 14

11) 2, 6, 5, 15

12) 7, 2, 14, 4

Similar Figures

Helpful Hints	– Two or more figures are similar if the corresponding angles are equal, and the corresponding sides are in proportion.	**Example:** 3–4–5 triangle is similar to a 6–8–10 triangle

Each pair of figures is similar. Find the missing side.

1)

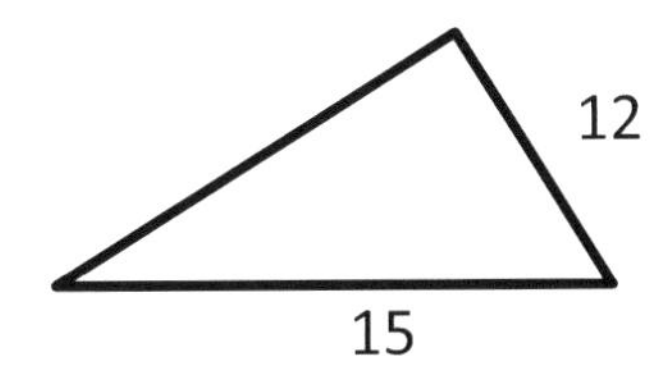

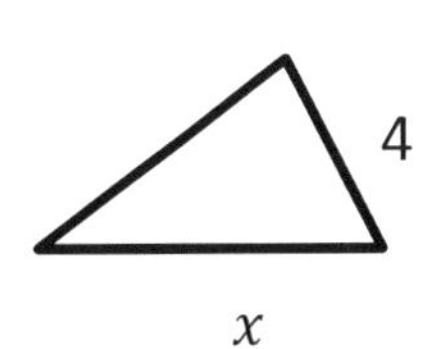

2)

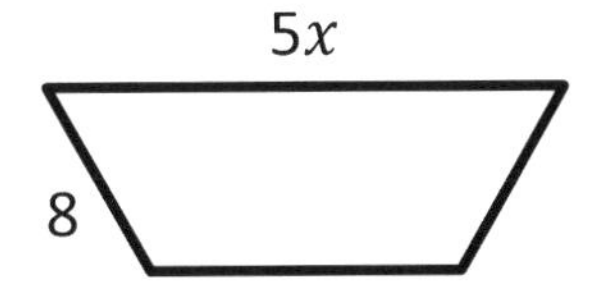

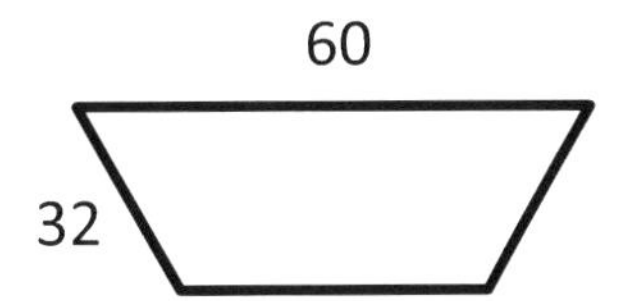

3)

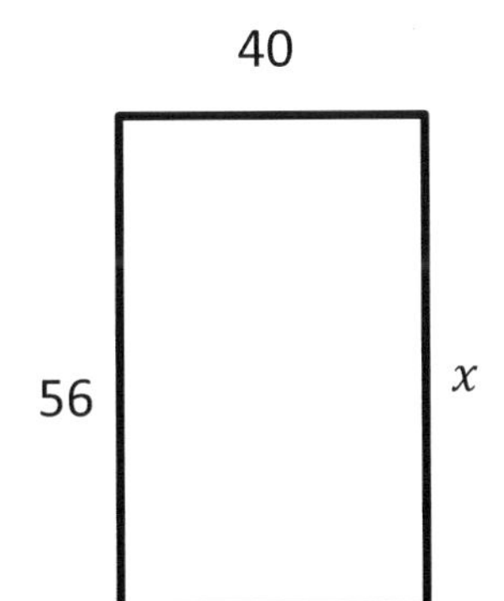

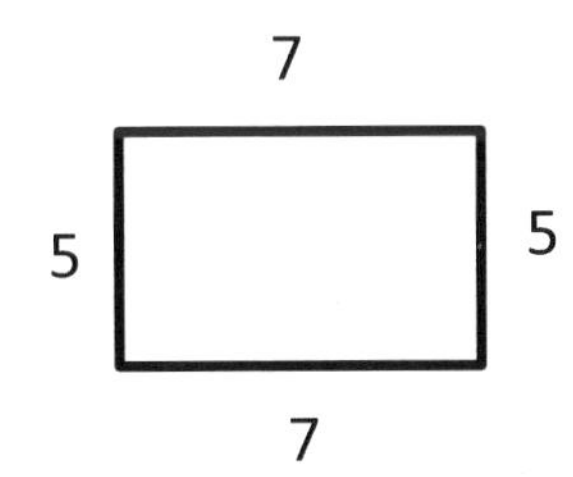

Simple Interest

Helpful Hints

Simple Interest: The charge for borrowing money or the return for lending it.
Interest = principal x rate x time

$$I = prt$$

Example:

$450 at 7% for 8 years.

$$I = prt$$

$$I = 450 \times 0.07 \times 8 = \$252$$

Use simple interest to find the ending balance.

1) $1,300 at 5% for 6 years.

2) $5,400 at 7.5% for 6 months.

3) $25,600 at 9.2% for 5 years

4) $24,000 at 8.5% for 9 years.

5) $450 at 7% for 8 years.

6) $54,200 at 8% for 5 years.

7) $240 interest is earned on a principal of $1500 at a simple interest rate of 4% interest per year. For how many years was the principal invested?

8) A new car, valued at $28,000, depreciates at 9% per year from original price. Find the value of the car 3 years after purchase.

9) Sara puts $2,000 into an investment yielding 5% annual simple interest; she left the money in for five years. How much interest does Sara get at the end of those five years?

Ratio and Rates Word Problems

Helpful Hints	To solve a ratio or a rate word problem, create a proportion and use cross multiplication method!	**Example:** $\frac{x}{4}=\frac{8}{16}$ $16x = 4 \times 8$ $x = 2$

Solve.

1) In a party, 10 soft drinks are required for every 12 guests. If there are 252 guests, how many soft drinks is required?

2) In Jack's class, 18 of the students are tall and 10 are short. In Michael's class 54 students are tall and 30 students are short. Which class has a higher ratio of tall to short students?

3) Are these ratios equivalent?
12 cards to 72 animals, 11 marbles to 66 marbles

4) The price of 3 apples at the Quick Market is $1.44. The price of 5 of the same apples at Walmart is $2.50. Which place is the better buy?

5) The bakers at a Bakery can make 160 bagels in 4 hours. How many bagels can they bake in 16 hours? What is that rate per hour?

6) You can buy 5 cans of green beans at a supermarket for $3.40. How much does it cost to buy 35 cans of green beans?

Answers of Worksheets – Chapter 4

Writing Ratios

1) $\frac{120\ miles}{4\ gallons}$, 30 miles per gallon
2) $\frac{24\ dollars}{6\ books}$, 4.00 dollars per book
3) $\frac{200\ miles}{14\ gallons}$, 14.29 miles per gallon
4) $\frac{24"\ of\ snow}{8\ hours}$, 3 inches of snow per hour
5) $\frac{1}{10}$
6) $\frac{3}{7}$
7) $\frac{2}{3}$
8) $\frac{1}{4}$
9) $\frac{1}{6}$
10) $\frac{9}{13}$
11) $\frac{1}{4}$
12) $\frac{2}{5}$
13) $\frac{8}{75}$
14) $\frac{12}{43}$

Simplifying Ratios

1) 3 : 7
2) 1 : 2
3) 1 : 5
4) 7 : 9
5) 5 : 3
6) 7 : 3
7) 10 : 1
8) 3 : 2
9) 7 : 9
10) 2 : 5
11) 5 : 7
12) 7 : 9
13) 26 : 41
14) 1 : 3
15) 8 : 1
16) 1 : 2
17) 1 : 12
18) 1 : 2
19) 3 : 50
20) 1 : 10
21) 1: 6
22) 2 : 9
23) 17 : 20
24) 1 : 10

Create a Proportion

1) 1 : 3 = 2 : 6
2) 12 : 144 = 1 : 12
3) 2 : 4 = 8 : 16
4) 5 : 15 = 9 : 27
5) 7 : 42, 10 : 60
6) 7 : 21 = 8 : 24
7) 8 : 10 = 4 : 5
8) 2 : 3 = 8 : 12
9) 4 : 2 = 2 : 1
10) 7 : 3 = 14 : 6
11) 5 : 2 = 15 : 6
12) 7 : 2 = 14 : 4

Similar Figures

1) 5
2) 3
3) 56

Simple Interest

1) $1,690.00
2) $5,602.50
3) $37,376.00
4) $42,360.00
5) $702.00
6) $75,880.00
7) 4 years
8) $20,440
9) $500

Ratio and Rates Word Problems

1) 210
2) The ratio for both class is equal to 9 to 5.
3) Yes! Both ratios are 1 to 6
4) The price at the Quick Market is a better buy.
5) 640, the rate is 40 per hour.
6) $23.80

Chapter 5: Percent

Topics that you'll learn in this chapter:

- ✓ Percentage Calculations
- ✓ Converting Between Percent, Fractions, and Decimals
- ✓ Percent Problems
- ✓ Markup, Discount, and Tax

"The book of nature is written in the language of Mathematic" -Galileo

Percentage Calculations

Helpful Hints

- Use the following formula to find part, whole, or percent:

  $\text{part} = \frac{\text{percent}}{100} \times \text{whole}$

Example:

$\frac{20}{100} \times 100 = 20$

Calculate the percentages.

1) 50% of 25
2) 80% of 15
3) 30% of 34
4) 70% of 45
5) 10% of 0
6) 80% of 22
7) 65% of 8
8) 78% of 54
9) 50% of 80
10) 20% of 10
11) 40% of 40
12) 90% of 0
13) 20% of 70
14) 55% of 60
15) 80% of 10
16) 20% of 880
17) 70% of 100
18) 80% of 90

Solve.

19) 50 is what percentage of 75?
20) What percentage of 100 is 70
21) Find what percentage of 60 is 35.
22) 40 is what percentage of 80?

Converting Between Percent, Fractions, and Decimals

Helpful Hints	– To a percent: Move the decimal point 2 places to the right and add the % symbol. – Divide by 100 to convert a number from percent to decimal.	**Examples:** 30% = 0.3 0.24 = 24%

Converting fractions to decimals.

1) $\frac{50}{100}$

2) $\frac{38}{100}$

3) $\frac{15}{100}$

4) $\frac{80}{100}$

5) $\frac{7}{100}$

6) $\frac{35}{100}$

7) $\frac{90}{100}$

8) $\frac{20}{100}$

9) $\frac{7}{100}$

Write each decimal as a percent.

10) 0.5

11) 0.9

12) 0.002

13) 0.524

14) 0.1

15) 0.03

16) 3.63

17) 0.008

18) 4.78

Percent Problems

Helpful Hints		Example:
	Base = Part ÷ Percent Part = Percent × Base Percent = Part ÷ Base	2 is 10% of 20. 2 ÷ 0.10 = 20 2 = 0.10 × 20 0.10 = 2 ÷ 20

Solve each problem.

1) 51 is 340% of what?
2) 93% of what number is 97?
3) 27% of 142 is what number?
4) What percent of 125 is 29.3?
5) 60 is what percent of 126?
6) 67 is 67% of what?
7) 67 is 13% of what?
8) 41% of 78 is what?
9) 1 is what percent of 52.6?
10) What is 59% of 14 m?
11) What is 90% of 130 inches?
12) 16 inches is 35% of what?
13) 90% of 54.4 hours is what?
14) What percent of 33.5 is 21?

15) Liam scored 22 out of 30 marks in Algebra, 35 out of 40 marks in science and 89 out of 100 marks in mathematics. In which subject his percentage of marks in best?

16) Ella require 50% to pass. If she gets 280 marks and falls short by 20 marks, what were the maximum marks she could have got?

Markup, Discount, and Tax

Helpful Hints

- **Markup** = selling price – cost
  Markup rate = markup divided by the cost
- **Discount:**
  Multiply the regular price by the rate of discount

  Selling price =

  original price – discount
- **Tax:**
  To find tax, multiply the tax rate to the taxable amount (income, property value, etc.)

Example:

Original price of a microphone: $49.99, discount: 5%, tax: 5%

Selling price = 49.87

Find the selling price of each item.

1) Cost of a pen: $1.95, markup: 70%, discount: 40%, tax: 5%

2) Cost of a puppy: $349.99, markup: 41%, discount: 23%

3) Cost of a shirt: $14.95, markup: 25%, discount: 45%

4) Cost of an oil change: $21.95, markup: 95%

5) Cost of computer: $1,850.00, markup: 75%

Answers of Worksheets – Chapter 5

Percentage Calculations

1) 12.5
2) 12
3) 10.2
4) 31.5
5) 0
6) 17.6
7) 5.2
8) 42.12
9) 40
10) 2
11) 16
12) 0
13) 14
14) 33
15) 8
16) 176
17) 70
18) 72
19) 67%
20) 70%
21) 58%
22) 50%

Converting Between Percent, Fractions, and Decimals

1) 0.5
2) 0.38
3) 0.15
4) 0.8
5) 0.07
6) 0.35
7) 0.9
8) 0.2
9) 0.07
10) 50%
11) 90%
12) 0.2%
13) 52.4%
14) 10%
15) 3%
16) 363%
17) 0.8%
18) 478%

Percent Problems

1) 15
2) 104.3
3) 38.34
4) 23.44%
5) 47.6%
6) 100
7) 515.4
8) 31.98
9) 1.9%
10) 8.3 m
11) 117 inches
12) 45.7inches
13) 49 hours
14) 62.7%
15) Mathematics
16) 600

Markup, Discount, and Tax

1) $2.09
2) $379.98
3) $10.28
4) $36.22
5) $3,237.50

Chapter 6: Algebraic Expressions

Topics that you'll learn in this chapter:

- ✓ Expressions and Variables
- ✓ Simplifying Variable Expressions
- ✓ Simplifying Polynomial Expressions
- ✓ Translate Phrases into an Algebraic Statement
- ✓ The Distributive Property
- ✓ Evaluating One Variable
- ✓ Evaluating Two Variables
- ✓ Combining like Terms

Without mathematics, there's nothing you can do. Everything around you is mathematics. Everything around you is numbers." – Shakuntala Devi

Expressions and Variables

Helpful Hints	A variable is a letter that represents unknown numbers. A variable can be used in the same manner as all other numbers:		
	Addition	2 + a	2 plus a
	Subtraction	$y - 3$	y minus 3
	Division	$\frac{4}{x}$	4 divided by x
	Multiplication	5a	5 times a

Simplify each expression.

1) $x + 5x$,

use $x = 5$

2) $8(-3x + 9) + 6$,

use $x = 6$

3) $10x - 2x + 6 - 5$,

use $x = 5$

4) $2x - 3x - 9$,

use $x = 7$

5) $(-6)(-2x - 4y)$,

use $x = 1$, $y = 3$

6) $8x + 2 + 4y$,

use $x = 9$, $y = 2$

7) $(-6)(-8x - 9y)$,

use $x = 5$, $y = 5$

8) $6x + 5y$,

use $x = 7$, $y = 4$

Simplify each expression.

9) $5(-4 + 2x)$

10) $-3 - 5x - 6x + 9$

11) $6x - 3x - 8 + 10$

12) $(-8)(6x - 4) + 12$

13) $9(7x + 4) + 6x$

14) $(-9)(-5x + 2)$

Simplifying Variable Expressions

Helpful Hints

– Combine "like" terms. (values with same variable and same power)

– Use distributive property if necessary.

Distributive Property:

$a(b + c) = ab + ac$

Example:

$2x + 2(1 - 5x) =$

$2x + 2 - 10x = -8x + 2$

Simplify each expression.

1) $-2 - x^2 - 6x^2$

2) $3 + 10x^2 + 2$

3) $8x^2 + 6x + 7x^2$

4) $5x^2 - 12x^2 + 8x$

5) $2x^2 - 2x - x$

6) $(-6)(8x - 4)$

7) $4x + 6(2 - 5x)$

8) $10x + 8(10x - 6)$

9) $9(-2x - 6) - 5$

10) $3(x + 9)$

11) $7x + 3 - 3x$

12) $2.5x^2 \times (-8x)$

Simplify.

13) $-2(4 - 6x) - 3x,\ x = 1$

14) $2x + 8x,\ x = 2$

15) $9 - 2x + 5x + 2,\ x = 5$

16) $5(3x + 7),\ x = 3$

17) $2(3 - 2x) - 4,\ x = 6$

18) $5x + 3x - 8,\ x = 3$

19) $x - 7x,\ x = 8$

20) $5(-2 - 9x),\ x = 4$

Simplifying Polynomial Expressions

Helpful Hints

- In mathematics, a polynomial is an expression consisting of variables and coefficients that involves only the operations of addition, subtraction, multiplication, and non–negative integer exponents of variables.

$P(x) = a_0x^n + a_1x^{n-1} + \ldots + a_{n-2}2x^2 + a_{n-1}x + a_n$

Example:

An example of a polynomial of a single indeterminate x is

$x^2 - 4x + 7.$

An example for three variables is

$x^3 + 2xyz^2 - yz + 1$

Simplify each polynomial.

1) $4x^5 - 5x^6 + 15x^5 - 12x^6 + 3x^6$

2) $(-3x^5 + 12 - 4x) + (8x^4 + 5x + 5x^5)$

3) $10x^2 - 5x^4 + 14x^3 - 20x^4 + 15x^3 - 8x^4$

4) $-6x^2 + 5x^2 - 7x^3 + 12 + 22$

5) $12x^5 - 5x^3 + 8x^2 - 8x^5$

6) $5x^3 + 1 + x^2 - 2x - 10x$

7) $14x^2 - 6x^3 - 2x(4x^2 + 2x)$

8) $(4x^4 - 2x) - (4x - 2x^4)$

9) $(3x^2 + 1) - (4 + 2x^2)$

10) $(2x + 2) - (7x + 6)$

11) $(12x^3 + 4x^4) - (2x^4 - 6x^3)$

12) $(12 + 3x^3) + (6x^3 + 6)$

13) $(5x^2 - 3) + (2x^2 - 3x^3)$

14) $(23x^3 - 12x^2) - (2x^2 - 9x^3)$

15) $(4x - 3x^3) - (3x^3 + 4x)$

Translate Phrases into an Algebraic Statement

Helpful Hints

Translating key words and phrases into algebraic expressions:

Addition: plus, more than, the sum of, etc.

Subtraction: minus, less than, decreased, etc.

Multiplication: times, product, multiplied, etc.

Division: quotient, divided, ratio, etc.

Example:

eight more than a number is 20

$8 + x = 20$

Write an algebraic expression for each phrase.

1) A number increased by forty–two.
2) The sum of fifteen and a number
3) The difference between fifty–six and a number.
4) The quotient of thirty and a number.
5) Twice a number decreased by 25.
6) Four times the sum of a number and – 12.
7) A number divided by – 20.
8) The quotient of 60 and the product of a number and – 5.
9) Ten subtracted from a number.
10) The difference of six and a number.

The Distributive Property

Helpful Hints

Distributive Property:

$a\,(b + c) = ab + ac$

Example:

$3\,(4 + 3x)$

$= 12 + 9x$

Use the distributive property to simply each expression.

1) $-(-2-5x)$

2) $(-6x+2)(-1)$

3) $(-5)\,(x-2)$

4) $-(7-3x)$

5) $8\,(8+2x)$

6) $2\,(12+2x)$

7) $(-6x+8)\,4$

8) $(3-6x)(-7)$

9) $(-12)\,(2x+1)$

10) $(8-2x)\,9$

11) $(-2x)\,(-1+9x)-4x\,(4+5x)$

12) $3\,(-5x-3)+4(6-3x)$

13) $(-2)(x+4)-(2+3x)$

14) $(-4)(3x-2)+6\,(x+1)$

15) $(-5)(4x-1)+4\,(x+2)$

16) $(-3)(x+4)-(2+3x)$

Evaluating One Variable

Helpful Hints	– To evaluate one variable expression, find the variable and substitute a number for that variable. – Perform the arithmetic operations.	**Example:** $4x + 8, x = 6$ $4(6) + 8 = 24 + 8 = 32$

Simplify each algebraic expression.

1) $9 - x$, $x = 3$

2) $x + 2$, $x = 5$

3) $3x + 7$, $x = 6$

4) $x + (-5)$, $x = -2$

5) $3x + 6$, $x = 4$

6) $4x + 6$, $x = -1$

7) $10 + 2x - 6$, $x = 3$

8) $10 - 3x$, $x = 8$

9) $\frac{20}{x} - 3$, $x = 5$

10) $(-3) + \frac{x}{4} + 2x$, $x = 16$

11) $(-2) + \frac{x}{7}$, $x = 21$

12) $(-\frac{14}{x}) - 9 + 4x$, $x = 2$

13) $(-\frac{6}{x}) - 9 + 2x$, $x = 3$

14) $(-2) + \frac{x}{8}$, $x = 16$

Evaluating Two Variables

Helpful Hints

To evaluate an algebraic expression, substitute a number for each variable and perform the arithmetic operations.

Example:

$2x + 4y - 3 + 2,$

$x = 5, y = 3$

$2(5) + 4(3) - 3 + 2 = 10 + 12 - 3 + 2 = 21$

Simplify each algebraic expression.

1) $2x + 4y - 3 + 2,$

   $x = 5, y = 3$

2) $(-\frac{12}{x}) + 1 + 5y,$

   $x = 6, y = 8$

3) $(-4)(-2a - 2b),$

   $a = 5, b = 3$

4) $10 + 3x + 7 - 2y,$

   $x = 7, y = 6$

5) $9x + 2 - 4y,$

   $x = 7, y = 5$

6) $6 + 3(-2x - 3y),$

   $x = 9, y = 7$

7) $12x + y,$

   $x = 4, y = 8$

8) $x \times 4 \div y,$

   $x = 3, y = 2$

9) $2x + 14 + 4y,$

   $x = 6, y = 8$

10) $4a - (5 - b),$

   $a = 4, b = 6$

Combining like Terms

Helpful Hints		Example:
	– Terms are separated by "+" and "–" signs. – Like terms are terms with same variables and same powers. – Be sure to use the "+" or "–" that is in front of the coefficient.	$22x + 6 + 2x =$ $24x + 6$

Simplify each expression.

1) $5 + 2x - 8$

2) $(-2x + 6)\ 2$

3) $7 + 3x + 6x - 4$

4) $(-4) - (3)(5x + 8)$

5) $9x - 7x - 5$

6) $x - 12x$

7) $7\ (3x + 6) + 2x$

8) $(-11x) - 10x$

9) $3x - 12 - 5x$

10) $13 + 4x - 5$

11) $(-22x) + 8x$

12) $2\ (4 + 3x) - 7x$

13) $(-4x) - (6 - 14x)$

14) $5\ (6x - 1) + 12x$

15) $22x + 6 + 2x$

16) $(-13x) - 14x$

17) $(-6x) - 9 + 15x$

18) $(-6x) + 7x$

19) $(-5x) + 12 + 7x$

20) $(-3x) - 9 + 15x$

21) $20x - 19x$

Answers of Worksheets – Chapter 6

Expressions and Variables

1) 30
2) −66
3) 41
4) −16
5) 84
6) 82
7) 510
8) 62
9) $10x - 20$
10) $6 - 11x$
11) $3x + 2$
12) $44 - 48x$
13) $69x + 36$
14) $45x - 18$

Simplifying Variable Expressions

1) $-7x^2 - 2$
2) $10x^2 + 5$
3) $15x^2 + 6x$
4) $-7x^2 + 8x$
5) $2x^2 - 3x$
6) $-48x + 24$
7) $-26x + 12$
8) $90x - 48$
9) $-18x - 59$
10) $3x + 27$
11) $4x + 3$
12) $-20x^3$
13) 1
14) 20
15) 26
16) 80
17) −22
18) 16
19) −48
20) −190

Simplifying Polynomial Expressions

1) $-14x^6 + 19x^5$
2) $2x^5 + 8x^4 + x + 12$
3) $-33x^4 + 29x^3 + 10x^2$
4) $-7x^3 - x^2 + 34$
5) $4x^5 - 5x^3 + 8x^2$
6) $5x^3 + x^2 - 12x + 1$
7) $-14x^3 + 10x^2$
8) $6x^4 - 6x$
9) $x^2 - 3$
10) $-5x - 4$
11) $2x^4 + 18x^3$
12) $9x^3 + 18$
13) $-3x^3 + 7x^2 - 3$
14) $32x^3 - 14x^2$
15) $-6x^3$

Translate Phrases into an Algebraic Statement

1) $x + 42$
2) $15 + x$
3) $56 - x$
4) $30/x$
5) $2x - 25$
6) $4(x + (-12))$
7) $\frac{x}{-20}$
8) $\frac{60}{-5x}$
9) $x - 10$
10) $6 - x$

The Distributive Property

1) $5x + 2$
2) $6x - 2$
3) $-5x + 10$
4) $3x - 7$
5) $16x + 64$
6) $4x + 24$
7) $-24x + 32$
8) $42x - 21$
9) $-24x - 12$
10) $-18x + 72$
11) $-38x^2 - 14x$
12) $-27x + 15$
13) $-5x - 10$
14) $-6x + 14$
15) $-16x + 13$
16) $-6x - 14$

Evaluating One Variable

1) 6
2) 7
3) 25
4) −7
5) 18
6) 2
7) 10
8) −14
9) 1
10) 33
11) 1
12) −8
13) −5
14) 0

Evaluating Two Variables

1) 21
2) 39
3) 64
4) 26
5) 45
6) −111
7) 56
8) 6
9) 58
10) 17

Combining like Terms

1) $2x - 3$
2) $-4x + 12$
3) $9x + 3$
4) $-15x - 28$
5) $2x - 5$
6) $-11x$
7) $23x + 42$
8) $-21x$
9) $-2x - 12$
10) $4x + 8$
11) $-14x$
12) $-x + 8$
13) $10x - 6$
14) $42x - 5$
15) $24x + 6$
16) $-27x$
17) $9x - 9$
18) x
19) $2x + 12$
20) $12x - 9$
21) x

Chapter 7: Equations

Topics that you'll learn in this chapter:

- ✓ One– Step Equations
- ✓ Two– Step Equations
- ✓ Multi– Step Equations

"The study of mathematics, like the Nile, begins in minuteness but ends in magnificence."

– Charles Caleb Colton

One–Step Equations

Helpful Hints

- The values of two expressions on both sides of an equation are equal.
  $ax + b = c$
- You only need to perform one Math operation in order to solve the equation.

Example:

$-8x = 16$

$x = -2$

Solve each equation.

1) $x + 3 = 17$

2) $22 = (-8) + x$

3) $3x = (-30)$

4) $(-36) = (-6x)$

5) $(-6) = 4 + x$

6) $2 + x = (-2)$

7) $20x = (-220)$

8) $18 = x + 5$

9) $(-23) + x = (-19)$

10) $5x = (-45)$

11) $x - 12 = (-25)$

12) $x - 3 = (-12)$

13) $(-35) = x - 27$

14) $8 = 2x$

15) $(-6x) = 36$

16) $(-55) = (-5x)$

17) $x - 30 = 20$

18) $8x = 32$

19) $36 = (-4x)$

20) $4x = 68$

21) $30x = 300$

Two–Step Equations

Helpful Hints	– You only need to perform two math operations (add, subtract, multiply, or divide) to solve the equation. – Simplify using the inverse of addition or subtraction. – Simplify further by using the inverse of multiplication or division.	**Example:** $-2(x-1)=42$ $(x-1)=-21$ $x=-20$

Solve each equation.

1) $5(8+x)=20$

2) $(-7)(x-9)=42$

3) $(-12)(2x-3)=(-12)$

4) $6(1+x)=12$

5) $12(2x+4)=60$

6) $7(3x+2)=42$

7) $8(14+2x)=(-34)$

8) $(-15)(2x-4)=48$

9) $3(x+5)=12$

10) $\frac{3x-12}{6}=4$

11) $(-12)=\frac{x+15}{6}$

12) $110=(-5)(2x-6)$

13) $\frac{x}{8}-12=4$

14) $20=12+\frac{x}{4}$

15) $\frac{-24+x}{6}=(-12)$

16) $(-4)(5+2x)=(-100)$

17) $(-12x)+20=32$

18) $\frac{-2+6x}{4}=(-8)$

19) $\frac{x+6}{5}=(-5)$

20) $(-9)+\frac{x}{4}=(-15)$

Multi–Step Equations

Helpful Hints	– Combine "like" terms on one side. – Bring variables to one side by adding or subtracting. – Simplify using the inverse of addition or subtraction. – Simplify further by using the inverse of multiplication or division.	**Example:** $3x + 15 = -2x + 5$ Add 2x both sides $5x + 15 = +5$ Subtract 15 both sides $5x = -10$ Divide by 5 both sides $x = -2$

Solve each equation.

1) $-(2-2x)=10$

2) $-12=-(2x+8)$

3) $3x+15=(-2x)+5$

4) $-28=(-2x)-12x$

5) $2(1+2x)+2x=-118$

6) $3x-18=22+x-3+x$

7) $12-2x=(-32)-x+x$

8) $7-3x-3x=3-3x$

9) $6+10x+3x=(-30)+4x$

10) $(-3x)-8(-1+5x)=352$

11) $24=(-4x)-8+8$

12) $9=2x-7+6x$

13) $6(1+6x)=294$

14) $-10=(-4x)-6x$

15) $4x-2=(-7)+5x$

16) $5x-14=8x+4$

17) $40=-(4x-8)$

18) $(-18)-6x=6(1+3x)$

19) $x-5=-2(6+3x)$

20) $6=1-2x+5$

Answers of Worksheets – Chapter 7

One–Step Equations

1) 14
2) 30
3) – 10
4) 6
5) – 10
6) – 4
7) – 11
8) 13
9) 4
10) – 9
11) – 13
12) – 9
13) – 8
14) 4
15) – 6
16) 11
17) 50
18) 4
19) – 9
20) 17
21) 10

Two–Step Equations

1) – 4
2) 3
3) 2
4) 1
5) 0.5
6) $\frac{4}{3}$
7) $-\frac{73}{8}$
8) $\frac{2}{5}$
9) – 1
10) 12
11) – 87
12) – 8
13) 128
14) 32
15) – 48
16) 10
17) – 1
18) – 5
19) – 31
20) – 24

Multi–Step Equations

1) 6
2) 2
3) – 2
4) 2
5) – 20
6) 37
7) 22
8) $\frac{4}{3}$
9) – 4
10) – 8
11) – 6
12) 2
13) 8
14) 1
15) 5
16) – 6
17) – 8
18) – 1
19) – 1
20) 0

Chapter 8: Inequalities

Topics that you'll learn in this chapter:

- ✓ Graphing Single– Variable Inequalities
- ✓ One– Step Inequalities
- ✓ Two– Step Inequalities
- ✓ Multi– Step Inequalities

Without mathematics, there's nothing you can do. Everything around you is mathematics. Everything around you is numbers." – Shakuntala Devi

Graphing Single–Variable Inequalities

Helpful Hints	– Isolate the variable. – Find the value of the inequality on the number line. – For less than or greater than draw open circle on the value of the variable. – If there is an equal sign too, then use filled circle. – Draw a line to the right direction.

Draw a graph for each inequality.

1) $-2 > x$

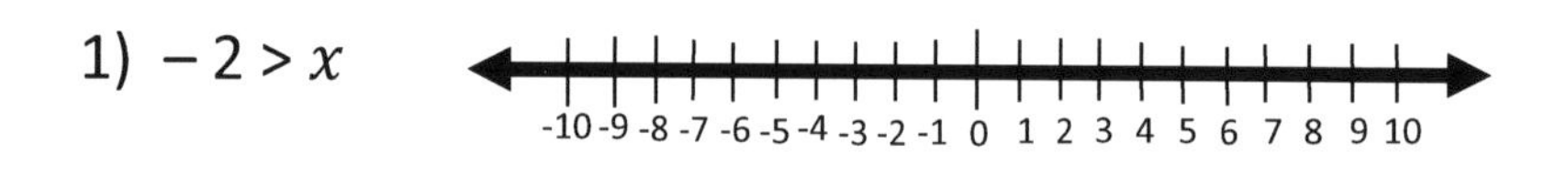

2) $5 \leq -x$

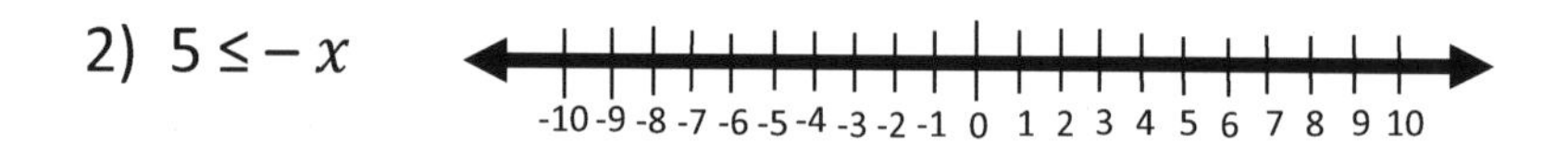

3) $x > 7$

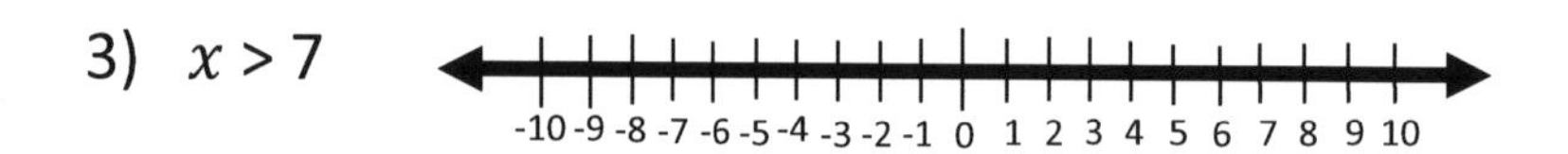

4) $-x > 1.5$

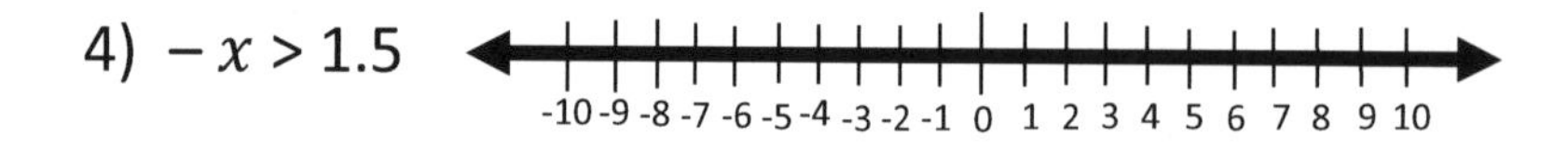

One–Step Inequalities

Helpful Hints	– Isolate the variable. – For dividing both sides by negative numbers, flip the direction of the inequality sign.	**Example:** $x + 4 \geq 11$ $x \geq 7$

Solve each inequality and graph it.

1) $x + 9 \geq 11$

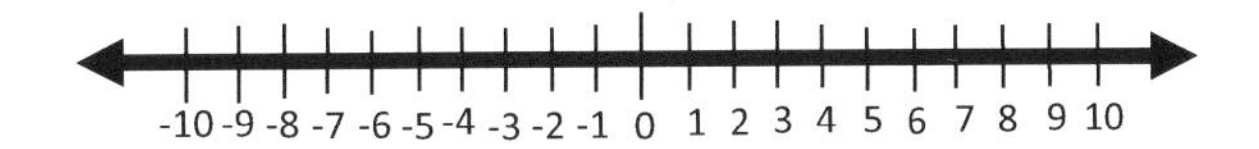

2) $x - 4 \leq 2$

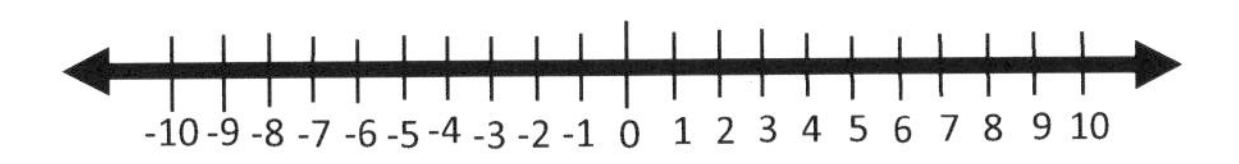

3) $6x \geq 36$

4) $7 + x < 16$

5) $x + 8 \leq 1$

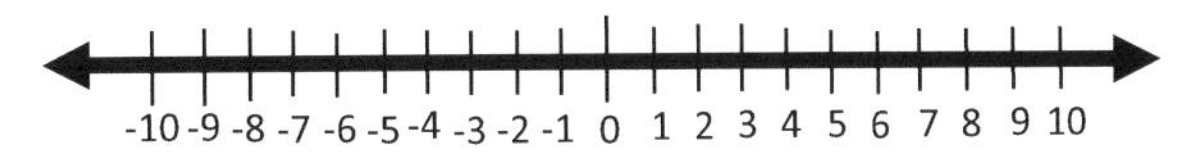

6) $3x > 12$

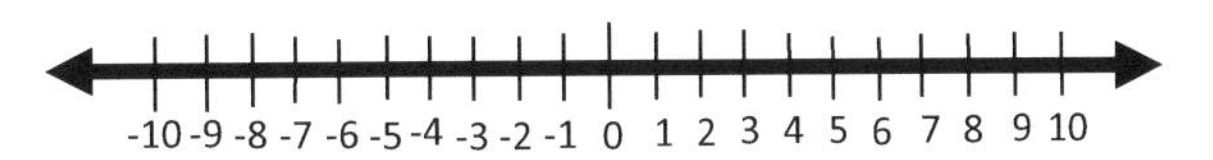

7) $3x < 24$

Two–Step Inequalities

Helpful Hints

– Isolate the variable.

– For dividing both sides by negative numbers, flip the direction of the of the inequality sign.

– Simplify using the inverse of addition or subtraction.

– Simplify further by using the inverse of multiplication or division.

Example:

$2x + 9 \geq 11$

$2x \geq 2$

$x \geq 1$

Solve each inequality and graph it.

1) $3x - 4 \leq 5$

2) $2x - 2 \leq 6$

3) $4x - 4 \leq 8$

4) $3x + 6 \geq 12$

5) $6x - 5 \geq 19$

6) $2x - 4 \leq 6$

7) $8x - 4 \leq 4$

8) $6x + 4 \leq 10$

9) $5x + 4 \leq 9$

10) $7x - 4 \leq 3$

11) $4x - 19 < 19$

12) $2x - 3 < 21$

13) $7 + 4x \geq 19$

14) $9 + 4x < 21$

15) $3 + 2x \geq 19$

16) $6 + 4x < 22$

Multi–Step Inequalities

Helpful Hints		Example:
	– Isolate the variable.	$\frac{7x+1}{3} \geq 5$
	– Simplify using the inverse of addition or subtraction.	$7x + 1 \geq 15$
	– Simplify further by using the inverse of multiplication or division.	$7x \geq 14$
		$x \geq 7$

Solve each inequality.

1) $\frac{9x}{7} - 7 < 2$

2) $\frac{4x+8}{2} \leq 12$

3) $\frac{3x-8}{7} > 1$

4) $-3\,(x-7) > 21$

5) $4 + \frac{x}{3} < 7$

6) $\frac{2x+6}{4} \leq 10$

Answers of Worksheets – Chapter 8

Graphing Single–Variable Inequalities

1) $-2 > x$

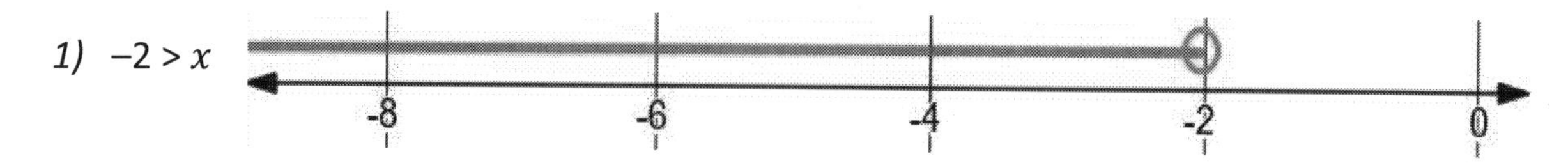

2) $x \leq -5$

3) $x > 7$

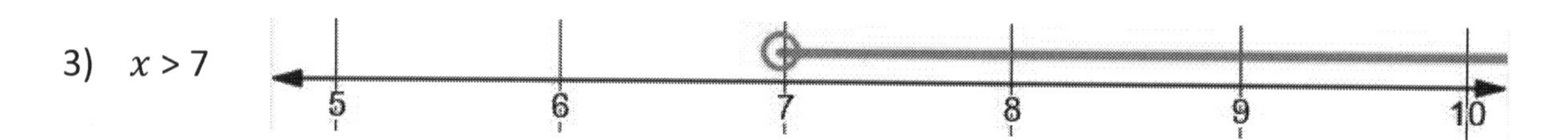

4) $-1.5 > x$

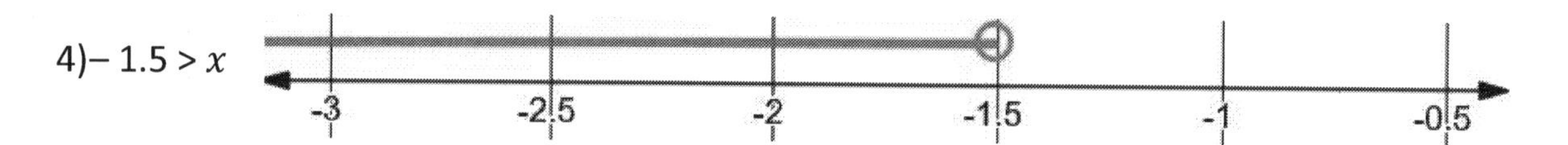

One–Step Inequalities

1)

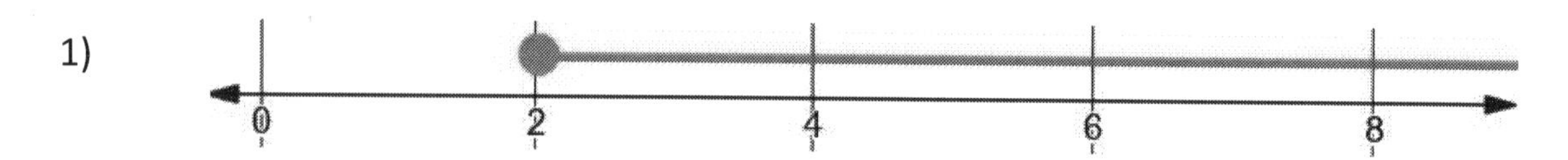

2)

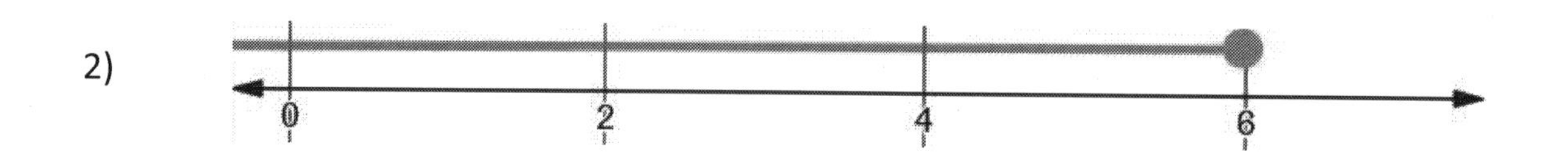

3)

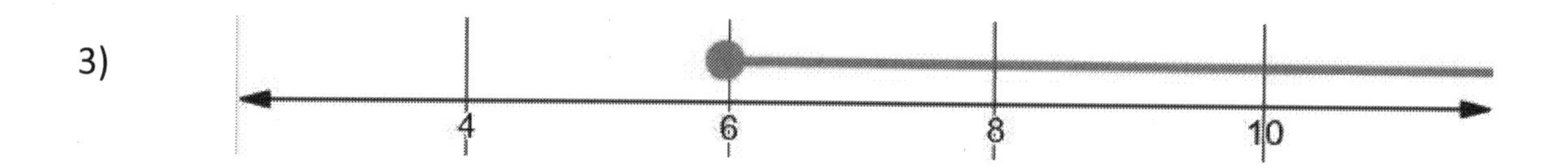

4)

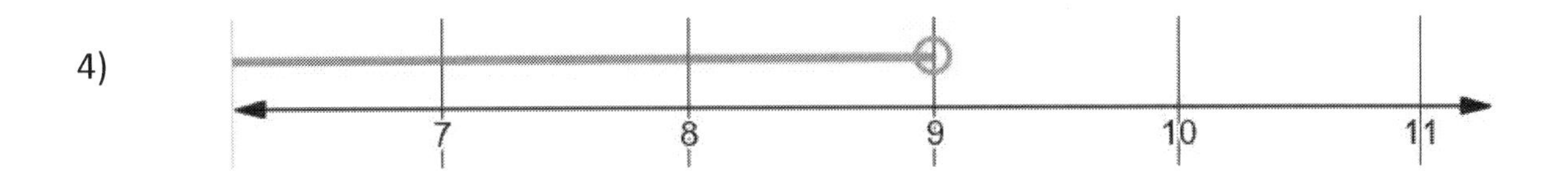

5)

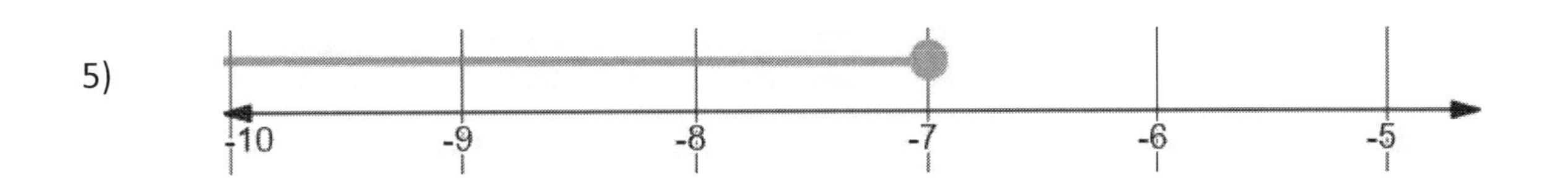

6)

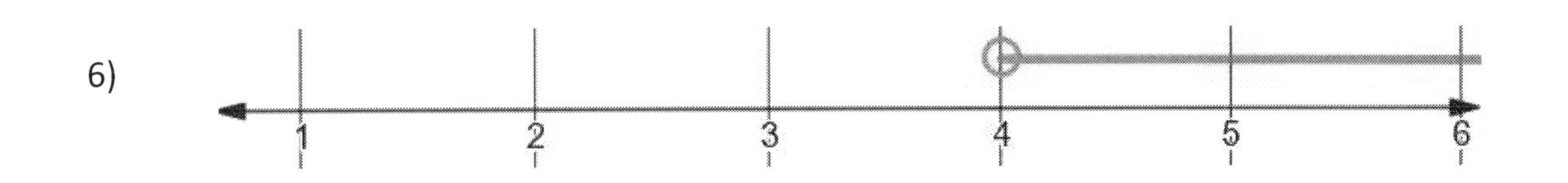

7)

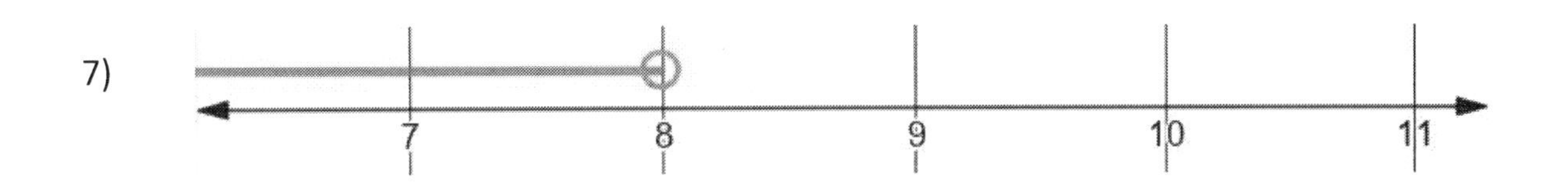

Two–Step inequalities

1) $x \leq 3$

2) $x \leq 4$

3) $x \leq 3$

4) $x \geq 2$

5) $x \geq 4$

6) $x \leq 5$

7) $x \leq 1$

8) $x \leq 1$

9) $x \leq 1$

10) $x \leq 1$

11) $x < 9.5$

12) $x < 12$

13) $x \geq 3$

14) $x < 3$

15) $x \geq 8$

16) $x < 4$

Multi–Step inequalities

1) $x < 7$

2) $x \leq 4$

3) $x > 5$

4) $x < 0$

5) $x < 9$

6) $x \leq 17$

Chapter 9: Linear Functions

Topics that you'll learn in this chapter:

- ✓ Finding Slope
- ✓ Graphing Lines Using Slope– Intercept Form
- ✓ Graphing Lines Using Standard Form
- ✓ Writing Linear Equations
- ✓ Graphing Linear Inequalities
- ✓ Finding Midpoint
- ✓ Finding Distance of Two Points

"Sometimes the questions are complicated and the answers are simple." – Dr. Seuss

Finding Slope

Helpful Hints	**Slope of a line:** $\frac{y_2 - y_1}{x_2 - x_1} = \frac{rise}{run}$	**Example:** (2, – 10), (3, 6) slope = 16

Find the slope of the line through each pair of points.

1) (1, 1), (3, 5)

2) (4, – 6), (– 3, – 8)

3) (7, – 12), (5, 10)

4) (19, 3), (20, 3)

5) (15, 8), (– 17, 9)

6) (6, – 12), (15, – 3)

7) (3, 1), (7, – 5)

8) (3, – 2), (– 7, 8)

9) (15, – 3), (– 9, 5)

10) (– 4, 7), (– 6, – 4)

11) (6, – 8), (– 11, – 7)

12) (– 6, 13), (17, – 9)

13) (– 10, – 2), (– 6, – 5)

14) (4, 5), (– 4, 10)

15) (– 3, 1), (– 17, 2)

16) (7, 0), (– 13, – 11)

17) (17, – 13), (17, 8)

18) (12, 2), (– 7, 5)

Graphing Lines Using Slope–Intercept Form

Helpful Hints

Slope–intercept form: given the slope m and the y–intercept b, then the equation of the line is:

$y = mx + b.$

Example:

$y = 8x - 3$

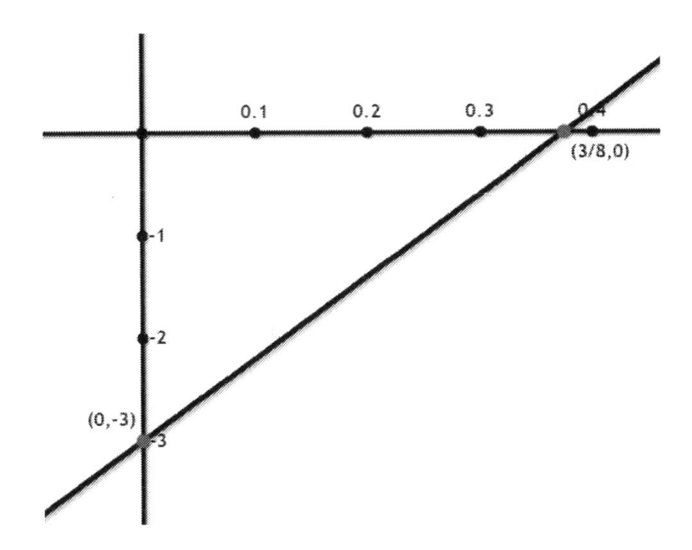

Sketch the graph of each line.

1) $y = \frac{1}{2}x - 4$

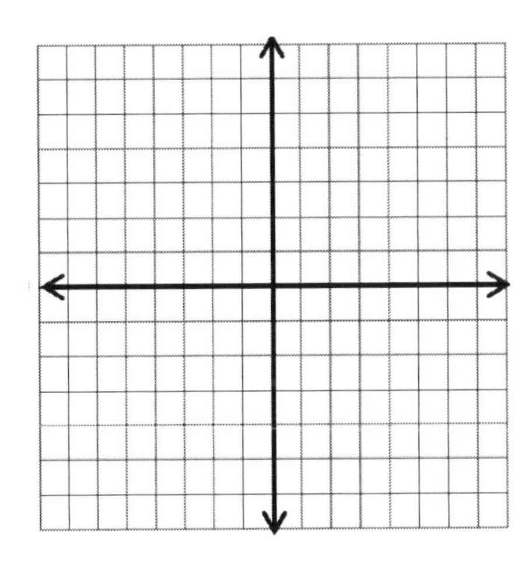

2) $y = 2x$

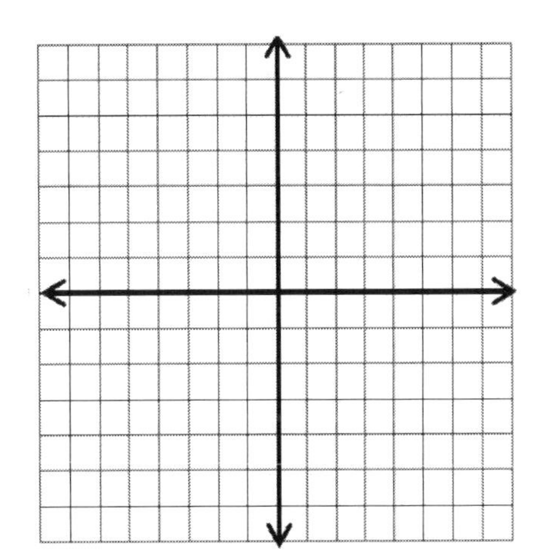

Graphing Lines Using Standard Form

Helpful Hints	– Find the –intercept of the line by putting zero for y. – Find the y–intercept of the line by putting zero for the x. – Connect these two points.

Example:

$x + 4y = 12$

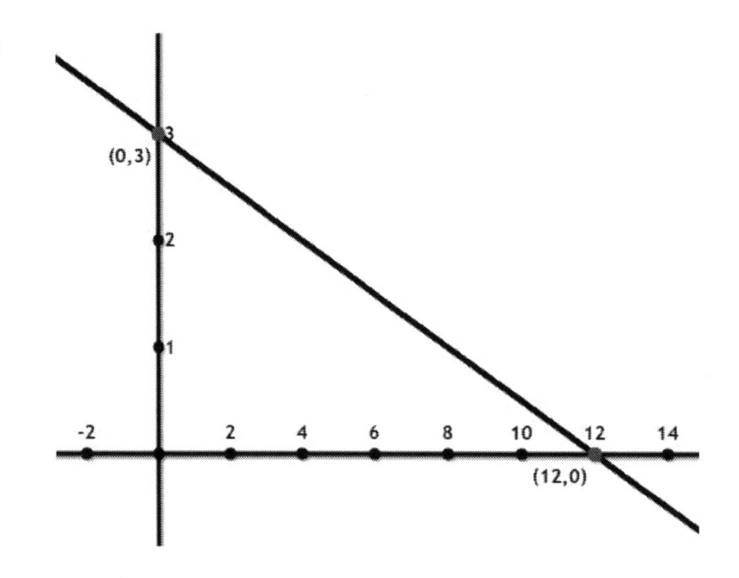

Sketch the graph of each line.

1) $2x - y = 4$

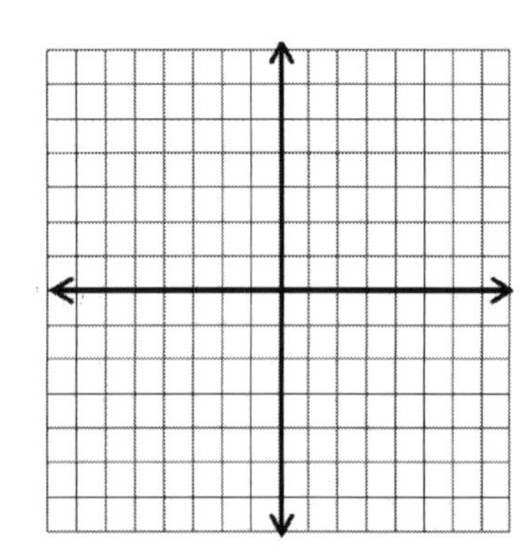

2) $x + y = 2$

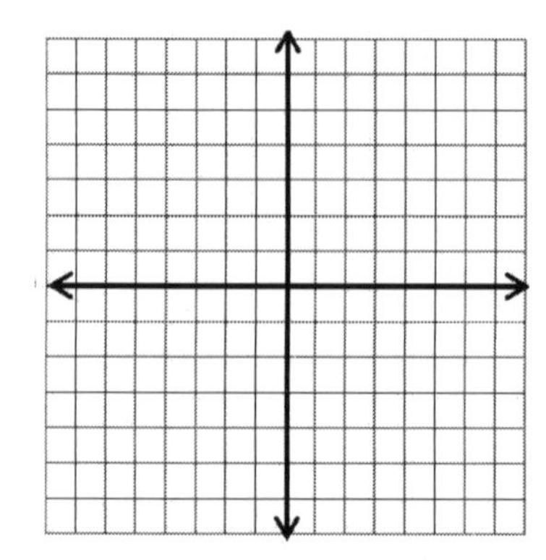

Writing Linear Equations

Helpful Hints

The equation of a line:

$$y = mx + b$$

1– Identify the slope.

2– Find the y–intercept. This can be done by substituting the slope and the coordinates of a point (x, y) on the line.

Example:

through:

(– 4, – 2), (– 3, 5)

$y = 7x + 26$

Write the slope–intercept form of the equation of the line through the given points.

1) through: (– 4, – 2), (– 3, 5)

2) through: (5, 4), (– 4, 3)

3) through: (0, – 2), (– 5, 3)

4) through: (– 1, 1), (– 2, 6)

5) through: (0, 3), (– 4, – 1)

6) through: (0, 2), (1, – 3)

7) through: (0, – 5), (4, 3)

8) through: (– 1, 4), (0, 4)

9) through: (2, – 3), (3, – 5)

10) through: (2, 5), (– 1, – 4)

11) through: (1, – 3), (– 3, 1)

12) through: (3, 3), (1, – 5)

13) through: (4, 4), (3, – 5)

14) through: (0, 3), (1, 1)

15) through: (5, 5), (2, – 3)

16) through: (– 2, – 2), (2, – 5)

17) through: (– 3, – 2), (1, – 1)

18) through: (–2, 1), (6, 5)

Graphing Linear Inequalities

Helpful Hints	1– First, graph the "equals" line. 2– Choose a testing point. (it can be any point on both sides of the line.) 3– Put the value of (x, y) of that point in the inequality. If that works, that part of the line is the solution. If the values don't work, then the other part of the line is the solution.

Sketch the graph of each linear inequality.

1) $y < -4x + 2$

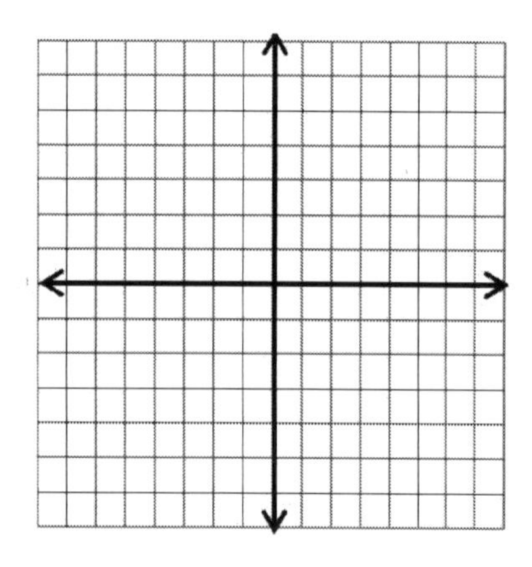

2) $2x + y < -4$

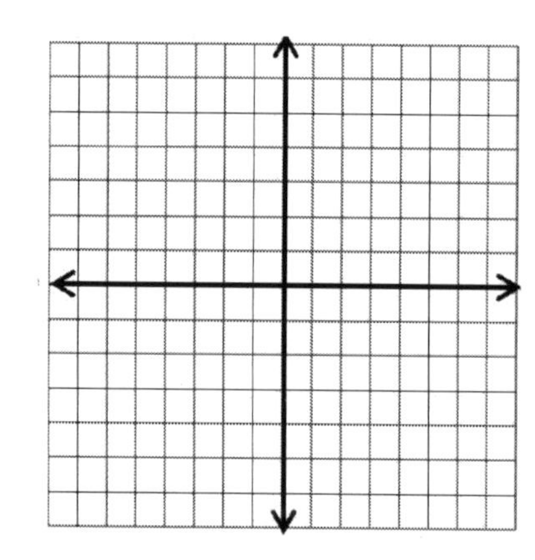

4) $x - 3y < -5$

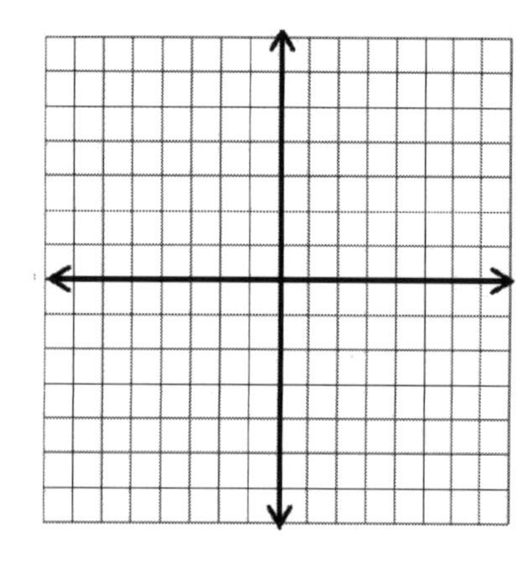

5) $6x - 2y \geq 8$

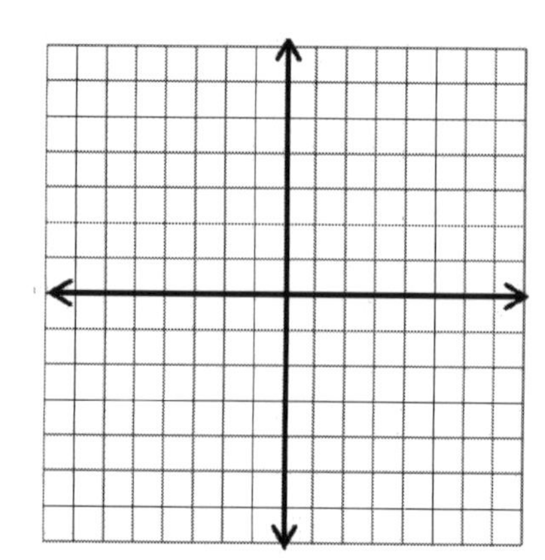

Finding Midpoint

Helpful Hints

Midpoint of the segment AB:

$$M\left(\frac{x_1+x_2}{2}, \frac{y_1+y_2}{2}\right)$$

Example:

(3, 9), (− 1, 6)

M (1, 7.5)

Find the midpoint of the line segment with the given endpoints.

1) (2, − 2), (3, − 5)

2) (0, 2), (− 2, − 6)

3) (7, 4), (9, − 1)

4) (4, − 5), (0, 8)

5) (1, − 2), (1, − 6)

6) (− 2, − 3), (3, − 6)

7) (7, 0), (− 7, 5)

8) (− 2, 6), (− 3, − 2)

9) (− 1, 1), (5, − 5)

10) (2.3, − 1.3), (− 2.2, − 0.5)

11) (4.1, 6.32), (4, 5.6)

12) (2, − 1), (− 6, 0)

13) (− 4, 4), (5, − 1)

14) (− 2, − 3), (− 6, 5)

15) $(\frac{1}{2}, 1)$, (2, 4)

16) (− 2, − 2), (6, 5)

Finding Distance of Two Points

Helpful Hints	Distance from A to B:	Example:
	$d = \sqrt{(x_1 - x_2)^2 + (y_1 - y_2)^2}$	(– 1, 2), (– 1, – 7) Distance = 9

Find the distance between each pair of points.

1) (2, –1), (1, – 1)

2) (6, 4), (– 1, 3)

3) (– 8, – 5), (– 6, 1)

4) (– 6, – 10), (– 2, – 10)

5) (4, – 6), (– 3, 4)

6) (– 6, – 7), (– 2, – 8)

7) (5, 4), (8, 2)

8) (8, 4), (3, – 7)

9) (1, 3), (5, 7)

10) (4, 2), (– 7, 1)

11) (– 3, – 4), (– 7, – 2)

12) (– 7, – 2), (6, 9)

13) (10, 0), (0, 4)

14) (– 3, 2), (5, 0)

15) (– 5, 6), (8, –4)

16) (3, – 5), (– 8, – 4)

17) (0, 8), (4, 10)

18) (6, 4), (– 5, – 1)

Answers of Worksheets – Chapter 9

Finding Slope

1) 2

2) $\frac{2}{7}$

3) -11

4) 0

5) $-\frac{1}{32}$

6) 1

7) $-\frac{3}{2}$

8) -1

9) $-\frac{1}{3}$

10) $\frac{11}{2}$

11) $-\frac{1}{17}$

12) $-\frac{22}{23}$

13) $-\frac{3}{4}$

14) $-\frac{5}{8}$

15) $-\frac{1}{14}$

16) $\frac{11}{20}$

17) Undefined

18) $-\frac{3}{19}$

Graphing Lines Using Slope–Intercept Form

1)

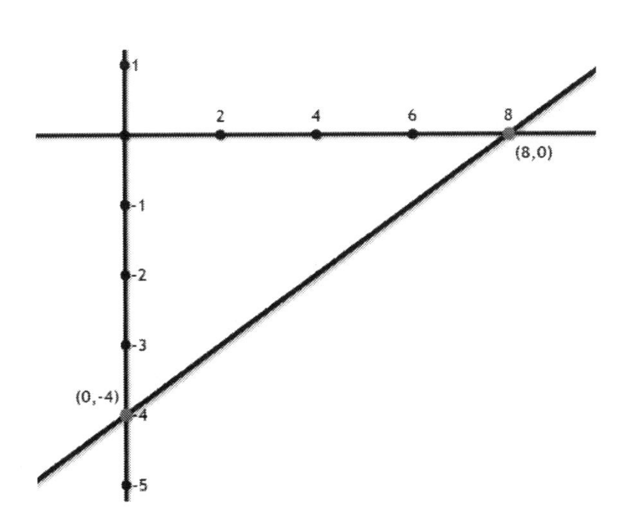

2)

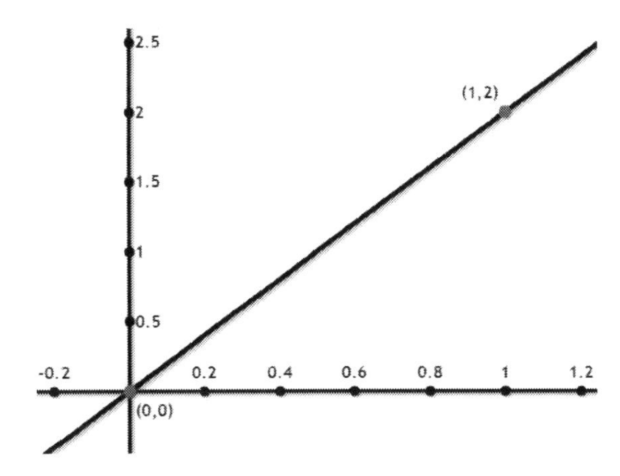

Graphing Lines Using Standard Form

1)

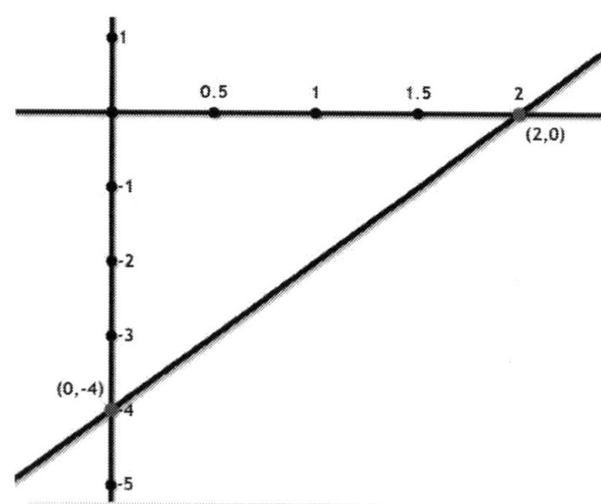

2)

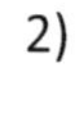

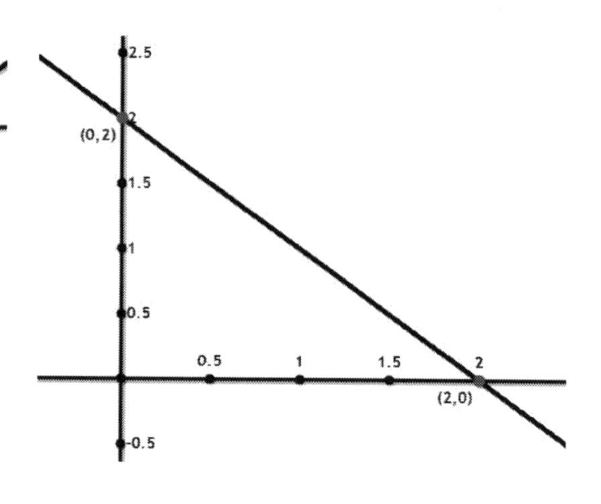

Writing Linear Equations

1) $y = 7x + 26$
2) $y = \frac{1}{9}x + \frac{31}{9}$
3) $y = -x - 2$
4) $y = -5x - 4$
5) $y = x + 3$
6) $y = -5x + 2$
7) $y = 2x - 5$
8) $y = 4$
9) $y = -2x + 1$
10) $y = 3x - 1$
11) $y = -x - 2$
12) $y = 4x - 9$
13) $y = 9x - 32$
14) $y = -2x + 3$
15) $y = \frac{8}{3}x - \frac{25}{3}$
16) $y = -\frac{3}{4}x - \frac{7}{2}$
17) $y = \frac{1}{4}x - \frac{5}{4}$
18) $y = -\frac{4}{3}x + \frac{19}{3}$

Graphing Linear Inequalities

1)

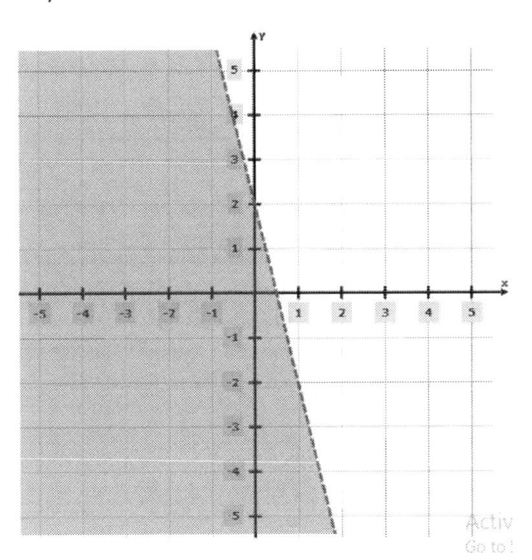

2)

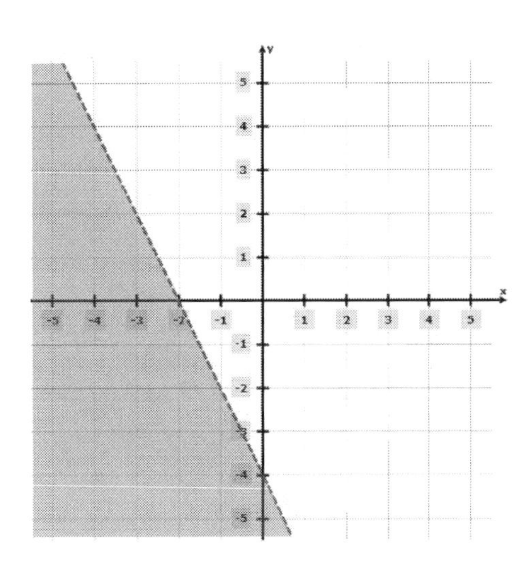

4)

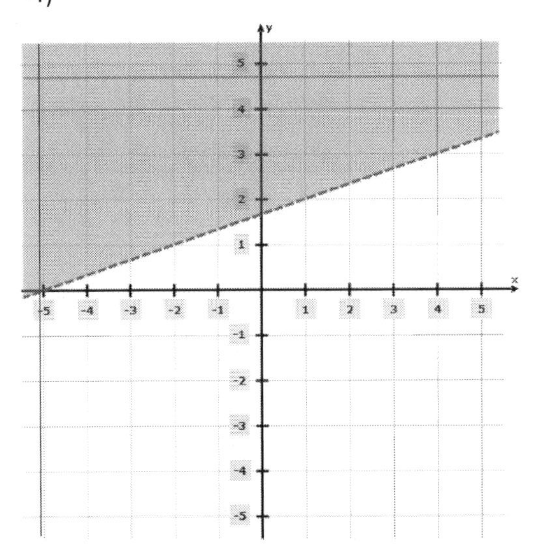

5)

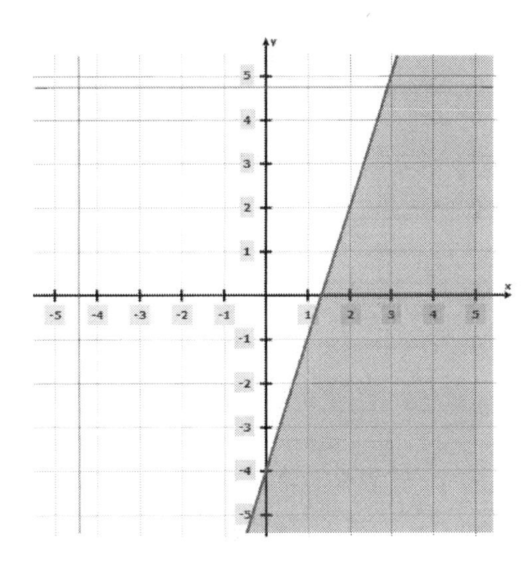

Finding Midpoint

1) (2.5, –3.5)
2) (–1, –2)
3) (8, 1.5)
4) (2, 1.5)
5) (1, –4)
6) (0.5, –4.5)
7) (0, 2.5)
8) (–2.5, 2)
9) (2, –2)
10) (0.05, –0.9)
11) (4.05, 5.96)
12) (–2, – 0.5)
13) $(\frac{1}{2}, 1\frac{1}{2})$
14) (–4, 1)
15) (1.25,2.5)
16) $(2, \frac{3}{2})$

Finding Distance of Two Points

1) 1
2) 7.1
3) 6.32
4) 4
5) 12.21
6) 4.12
7) 3.61
8) 12.1
9) 5.66
10) 11.04
11) 4.47
12) 17.03
13) 10.77
14) 8.25
15) 16.4
16) 10.3
17) 4.47
18) 12.1

Chapter 10: Polynomials

Topics that you'll learn in this chapter:

- ✓ Classifying Polynomials
- ✓ Writing Polynomials in Standard Form
- ✓ Simplifying Polynomials
- ✓ Adding and Subtracting Polynomials
- ✓ Multiplying Monomials
- ✓ Multiplying and Dividing Monomials
- ✓ Multiplying a Polynomial and a Monomial
- ✓ Multiplying Binomials
- ✓ Factoring Trinomials
- ✓ Operations with Polynomials

Mathematics – the unshaken Foundation of Sciences, and the plentiful Fountain of Advantage to human affairs. — Isaac Barrow

Classifying Polynomials

Helpful Hints

Name	Degree	Example
constant	0	4
linear	1	$2x$
quadratic	2	$x^2 + 5x + 6$
cubic	3	$x^3 - x^2 + 4x + 8$
quartic	4	$x^4 + 3x^3 - x^2 + 2x + 6$
quantic	5	$x^5 - 2x^4 + x^3 - x^2 + x + 10$

Name each polynomial by degree and number of terms.

1) x

2) $-5x^4$

3) $7x - 4$

4) -6

5) $8x + 1$

6) $9x^2 - 8x^3$

7) $2x^5$

8) $10 + 8x$

9) $5x^2 - 6x$

10) $-7x^7 + 7x^4$

11) $-8x^4 + 5x^3 - 2x^2 - 8x$

12) $4x - 9x^2 + 4x^3 - 5x^4$

13) $4x^6 + 5x^5 + x^4$

14) $-4 - 2x^2 + 8x$

15) $9x^6 - 8$

16) $7x^5 + 10x^4 - 3x + 10x^7$

17) $4x^6 - 3x^2 - 8x^4$

18) $-5x^4 + 10x - 10$

Writing Polynomials in Standard Form

Helpful Hints

A polynomial function $f(x)$ of degree n is of the form

$f(x) = a_n x^n + a_{n-1} x^{n-1} + \dots + a_1 x + a_0$

The first term is the one with the biggest power!

Example:

$2x^2 - 4x^3 - x =$

$-4x^3 + 2x^2 - x$

Write each polynomial in standard form.

1) $3x^2 - 5x^3$

2) $3 + 4x^3 - 3$

3) $2x^2 + 1x - 6x^3$

4) $9x - 7x$

5) $12 - 7x + 9x^4$

6) $5x^2 + 13x - 2x^3$

7) $-3 + 16x - 16x$

8) $3x(x + 4) - 2(x + 4)$

9) $(x + 5)(x - 2)$

10) $3x^2 + x + 12 - 5x^2 - 2x$

11) $12x^5 + 7x^3 - 3x^5 - 8x^3$

12) $3x(2x + 5 - 2x^2)$

13) $11x(x^5 + 2x^3)$

14) $(x + 6)(x + 3)$

15) $(x + 4)^2$

16) $(8x - 7)(3x + 2)$

17) $5x(3x^2 + 2x + 1)$

18) $7x(3 - x + 6x^3)$

Simplifying Polynomials

Helpful Hints		Example:
	1– Find "like" terms. (they have same variables with same power).	
	2– Add or Subtract "like" terms using PEMDAS operation.	$2x^5 - 3x^3 + 8x^2 - 2x^5 =$ $-3x^3 + 8x^2$

Simplify each expression.

1) $11 - 4x^2 + 3x^2 - 7x^3 + 3$

2) $2x^5 - x^3 + 8x^2 - 2x^5$

3) $(-5)(x^6 + 10) - 8(14 - x^6)$

4) $4(2x^2 + 4x^2 - 3x^3) + 6x^3 + 17$

5) $11 - 6x^2 + 5x^2 - 12x^3 + 22$

6) $2x^2 - 2x + 3x^3 + 12x - 22x$

7) $(3x - 8)(3x - 4)$

8) $(12x + 2y)^2$

9) $(12x^3 + 28x^2 + 10x + 4) \div (x + 2)$

10) $(2x + 12x^2 - 2) \div (2x + 1)$

11) $(2x^3 - 1) + (3x^3 - 2x^3)$

12) $(x - 5)(x - 3)$

13) $(3x + 8)(3x - 8)$

14) $(8x^2 - 3x) - (5x - 5 - 8x^2)$

Adding and Subtracting Polynomials

Helpful Hints

Adding polynomials is just a matter of combining like terms, with some order of operations considerations thrown in.

Be careful with the minus signs, and don't confuse addition and multiplication!

Example:

$(3x^3 - 1) - (4x^3 + 2)$

$= - x^3 - 3$

Simplify each expression.

1) $(2x^3 - 2) + (2x^3 + 2)$

2) $(4x^3 + 5) - (7 - 2x^3)$

3) $(4x^2 + 2x^3) - (2x^3 + 5)$

4) $(4x^2 - x) + (3x - 5x^2)$

5) $(7x + 9) - (3x + 9)$

6) $(4x^4 - 2x) - (6x - 2x^4)$

7) $(12x - 4x^3) - (8x^3 + 6x)$

8) $(2x^3 - 8x^2) - (5x^2 - 3x^3)$

9) $(2x^2 - 6) + (9x^2 - 4x^3)$

10) $(4x^3 + 3x^4) - (x^4 - 5x^3)$

11) $(- 12x^4 + 10x^5 + 2x^3) + (14x^3 + 23x^5 + 8x^4)$

12) $(13x^2 - 6x^5 - 2x) - (-10x^2 - 11x^5 + 9x)$

13) $(35 + 9x^5 - 3x^2) + (8x^4 + 3x^5) - (27 - 5x^4)$

14) $(3x^5 - 2x^3 - 4x) + (4x + 10x^4 - 23) + (x^2 - x^3 + 12)$

Multiplying Monomials

Helpful Hints	A monomial is a polynomial with just one term, like $2x$ or $7y$.	**Example:** $2u^3 \times (-3u)$ $= -6u^4$

Simplify each expression.

1) $2xy^2z \times 4z^2$

2) $4xy \times x^2y$

3) $4pq^3 \times (-2p^4q)$

4) $8s^4t^2 \times st^5$

5) $12p^3 \times (-3p^4)$

6) $-4p^2q^3r \times 6pq^2r^3$

7) $(-8a^4) \times (-12a^6b)$

8) $3u^4v^2 \times (-7u^2v^3)$

9) $4u^3 \times (-2u)$

10) $-6xy^2 \times 3x^2y$

11) $12y^2z^3 \times (-y^2z)$

12) $5a^2bc^2 \times 2abc^2$

Multiplying and Dividing Monomials

Helpful Hints

- When you divide two monomials you need to divide their coefficients and then divide their variables.
- In case of exponents with the same base, you need to subtract their powers.

Example:

$(-3x^2)(8x^4y^{12}) = -24x^6y^{12}$

$\frac{36x^5y^7}{4x^4y^5} = 9xy^2$

Simplify.

1) $(7x^4y^6)(4x^3y^4)$

2) $(15x^4)(3x^9)$

3) $(12x^2y^9)(7x^9y^{12})$

4) $\frac{80x^{12}y^9}{10x^6y^7}$

5) $\frac{95x^{18}y^7}{5x^9y^2}$

6) $\frac{200x^3y^8}{40x^3y^7}$

7) $\frac{-15x^{17}y^{13}}{3x^6y^9}$

8) $\frac{-64x^8y^{10}}{8x^3y^7}$

Multiplying a Polynomial and a Monomial

Helpful Hints

– When multiplying monomials, use the product rule for exponents.

– When multiplying a monomial by a polynomial, use the distributive property.

a × (b + c) = a × b + a × c

Example:

$2x\,(8x - 2) =$

$16x^2 - 4x$

Find each product.

1) $5\,(3x - 6y)$

2) $9x\,(2x + 4y)$

3) $8x\,(7x - 4)$

4) $12x\,(3x + 9)$

5) $11x\,(2x - 11y)$

6) $2x\,(6x - 6y)$

7) $3x\,(2x^2 - 3x + 8)$

8) $13x\,(4x + 8y)$

9) $20\,(2x^2 - 8x - 5)$

10) $3x\,(3x - 2)$

11) $6x^3\,(3x^2 - 2x + 2)$

12) $8x^2\,(3x^2 - 5xy + 7y^2)$

13) $2x^2\,(3x^2 - 5x + 12)$

14) $2x^3\,(2x^2 + 5x - 4)$

15) $5x\,(6x^2 - 5xy + 2y^2)$

16) $9\,(x^2 + xy - 8y^2)$

Multiplying Binomials

Helpful Hints	Use "FOIL". (First–Out–In–Last) $(x + a)(x + b) = x^2 + (b + a)x + ab$	**Example:** $(x + 2)(x - 3) =$ $x^2 - x - 6$

Multiply.

1) $(3x - 2)(4x + 2)$

2) $(2x - 5)(x + 7)$

3) $(x + 2)(x + 8)$

4) $(x^2 + 2)(x^2 - 2)$

5) $(x - 2)(x + 4)$

6) $(x - 8)(2x + 8)$

7) $(5x - 4)(3x + 3)$

8) $(x - 7)(x - 6)$

9) $(6x + 9)(4x + 9)$

10) $(2x - 6)(5x + 6)$

11) $(x - 7)(x + 7)$

12) $(x + 4)(4x - 8)$

13) $(6x - 4)(6x + 4)$

14) $(x - 7)(x + 2)$

15) $(x - 8)(x + 8)$

16) $(3x + 3)(3x - 4)$

17) $(x + 3)(x + 3)$

18) $(x + 4)(x + 6)$

Factoring Trinomials

Helpful Hints

"FOIL"

$(x + a)(x + b) = x^2 + (b + a)x + ab$

"Difference of Squares"

$a^2 - b^2 = (a + b)(a - b)$

$a^2 + 2ab + b^2 = (a + b)(a + b)$

$a^2 - 2ab + b^2 = (a - b)(a - b)$

"Reverse FOIL"

$x^2 + (b + a)x + ab = (x + a)(x + b)$

Example:

$x^2 + 5x + 6 =$

$(x + 2)(x + 3)$

Factor each trinomial.

1) $x^2 - 7x + 12$

2) $x^2 + 5x - 14$

3) $x^2 - 11x - 42$

4) $6x^2 + x - 12$

5) $x^2 - 17x + 30$

6) $x^2 + 8x + 15$

7) $3x^2 + 11x - 4$

8) $x^2 - 6x - 27$

9) $10x^2 + 33x - 7$

10) $x^2 + 24x + 144$

11) $49x^2 + 28xy + 4y^2$

12) $16x^2 - 40x + 25$

13) $x^2 - 10x + 25$

14) $25x^2 - 20x + 4$

15) $x^3 + 6x^2y^2 + 9xy^3$

16) $9x^2 + 24x + 16$

17) $x^2 - 8x + 16$

18) $x^2 + 121 + 22x$

Operations with Polynomials

Helpful Hints	– When multiplying a monomial by a polynomial, use the distributive property.	**Example:**
	a × (b + c) = a × b + a × c	$5\,(6x - 1) =$
		$30x - 5$

Find each product.

1) $3x^2\,(6x - 5)$

2) $5x^2\,(7x - 2)$

3) $-3\,(8x - 3)$

4) $6x^3\,(-3x + 4)$

5) $9\,(6x + 2)$

6) $8\,(3x + 7)$

7) $5\,(6x - 1)$

8) $-7x^4\,(2x - 4)$

9) $8\,(x^2 + 2x - 3)$

10) $4\,(4x^2 - 2x + 1)$

11) $2\,(3x^2 + 2x - 2)$

12) $8x\,(5x^2 + 3x + 8)$

13) $(9x + 1)\,(3x - 1)$

14) $(4x + 5)\,(6x - 5)$

15) $(7x + 3)\,(5x - 6)$

16) $(3x - 4)\,(3x + 8)$

Answers of Worksheets – Chapter 10

Classifying Polynomials

1) Linear monomial
2) Quartic monomial
3) Linear binomial
4) Constant monomial
5) Linear binomial
6) Cubic binomial
7) Quantic monomial
8) Linear binomial
9) Quadratic binomial
10) Seventh degree binomial
11) Quartic polynomial with four terms
12) Quartic polynomial with four terms
13) Sixth degree trinomial
14) Quadratic trinomial
15) Sixth degree binomial
16) Seventh degree polynomial with four terms
17) Sixth degree trinomial
18) Quartic trinomial

Writing Polynomials in Standard Form

1) $-5x^3 + 3x^2$
2) $4x^3$
3) $-6x^3 + 2x^2 + x$
4) $2x$
5) $9x^4 - 7x + 12$
6) $-2x^3 + 5x^2 + 13x$
7) -3
8) $3x^2 + 10x - 8$
9) $x^2 + 3x - 10$
10) $-2x^2 - x + 12$
11) $9x^5 - x^3$
12) $-6x^3 + 6x^2 + 15x$
13) $11x^6 + 22x^4$
14) $x^2 + 9x + 18$
15) $x^2 + 8x + 16$
16) $24x^2 - 5x - 14$
17) $15x^3 + 10x^2 + 5x$
18) $42x^4 - 7x^2 + 21x$

Simplifying Polynomials

1) $-7x^3 - x^2 + 14$
2) $-x^3 + 8x^2$
3) $3x^6 - 162$
4) $-6x^3 + 24x^2 + 17$
5) $-12x^3 - x^2 + 33$
6) $3x^3 + 2x^2 - 12x$

7) $9x^2 - 36x + 32$

8) $144x^2 + 48xy + 4y^2$

9) $12x^2 + 4x + 2$

10) $6x - 1$

11) $3x^3 - 1$

12) $x^2 - 8x + 15$

13) $9x^2 - 64$

14) $16x^2 - 8x + 5$

Adding and Subtracting Polynomials

1) $4x^3$

2) $6x^3 - 2$

3) $4x^2 - 5$

4) $-x^2 + 2x$

5) $4x$

6) $6x^4 - 8x$

7) $-12x^3 + 6x$

8) $5x^3 - 13x^2$

9) $-4x^3 + 11x^2 - 6$

10) $2x^4 + 9x^3$

11) $33x^5 - 4x^4 + 16x^3$

12) $5x^5 + 23x^2 - 11x$

13) $12x^5 + 13x^4 - 3x^2 + 8$

14) $3x^5 + 10x^4 - 3x^3 + x^2 - 11$

Multiplying Monomials

1) $8xy^2z^3$

2) $4x^3y^2$

3) $-8p^5q^4$

4) $8s^5t^7$

5) $-36p^7$

6) $-24p^3q^5r^4$

7) $96a^{10}b$

8) $-21u^6v^5$

9) $-8u^4$

10) $-18x^3y^3$

11) $-12y^4z^4$

12) $10a^3b^2c^4$

Multiplying and Dividing Monomials

1) $28x^7y^{10}$
2) $45x^{13}$
3) $84x^{11}y^{21}$
4) $8x^6y^2$
5) $19x^9y^5$
6) $5y$
7) $-5x^{11}y^4$
8) $-8x^5y^3$

Multiplying a Polynomial and a Monomial

1) $15x - 30y$
2) $18x^2 + 36xy$
3) $56x^2 - 32x$
4) $36x^2 + 108x$
5) $22x^2 - 121xy$
6) $12x^2 - 12xy$
7) $6x^3 - 9x^2 + 24x$
8) $52x^2 + 104xy$
9) $40x^2 - 160x - 100$
10) $9x^2 - 6x$
11) $18x^5 - 12x^4 + 12x^3$
12) $24x^4 - 40x^3y + 56y^2x^2$
13) $6x^4 - 10x^3 + 24x^2$
14) $4x^5 + 10x^4 - 8x^3$
15) $30x^3 - 25x^2y + 10xy^2$
16) $9x^2 + 9xy - 72y^2$

Multiplying Binomials

1) $12x^2 - 2x - 4$
2) $2x^2 + 9x - 35$
3) $x^2 + 10x + 16$
4) $x^4 - 4$
5) $x^2 + 2x - 8$
6) $2x^2 - 8x - 64$
7) $15x^2 + 3x - 12$
8) $x^2 - 13x + 42$
9) $24x^2 + 90x + 81$
10) $10x^2 - 18x - 36$
11) $x^2 - 49$
12) $4x^2 + 8x - 32$
13) $36x^2 - 16$
14) $x^2 - 5x - 14$
15) $x^2 - 64$
16) $9x^2 - 3x - 12$
17) $x^2 + 6x + 9$
18) $x^2 + 10x + 24$

Factoring Trinomials

1) $(x - 3)(x - 4)$
2) $(x - 2)(x + 7)$
3) $(x + 3)(x - 14)$
4) $(2x + 3)(3x - 4)$
5) $(x - 15)(x - 2)$
6) $(x + 3)(x + 5)$
7) $(3x - 1)(x + 4)$
8) $(x - 9)(x + 3)$
9) $(5x - 1)(2x + 7)$
10) $(x + 12)(x + 12)$
11) $(7x + 2y)(7x + 2y)$
12) $(4x - 5)(4x - 5)$
13) $(x - 5)(x - 5)$
14) $(5x - 2)(5x - 2)$
15) $x(x^2 + 6xy^2 + 9y^3)$
16) $(3x + 4)(3x + 4)$
17) $(x - 4)(x - 4)$
18) $(x + 11)(x + 11)$

Operations with Polynomials

1) $18x^3 - 15x^2$
2) $35x^3 - 10x^2$
3) $-24x + 9$
4) $-18x^4 + 24x^3$
5) $54x + 18$
6) $24x + 56$
7) $30x - 5$
8) $-14x^5 + 28x^4$
9) $8x^2 + 16x - 24$
10) $16x^2 - 8x + 4$
11) $6x^2 + 4x - 4$
12) $40x^3 + 24x^2 + 64x$
13) $27x^2 - 6x - 1$
14) $24x^2 + 10x - 25$
15) $35x^2 - 27x - 18$
16) $9x^2 + 12x - 32$

Chapter 11: Systems of Equations

Topics that you'll learn in this chapter:

- ✓ Solving Systems of Equations by Substitution
- ✓ Solving Systems of Equations by Elimination
- ✓ Systems of Equations Word Problems

Mathematics is the door and key to the sciences. — Roger Bacon

Solving Systems of Equations by Substitution

Helpful Hints	Consider the system of equations $x - y = 1, -2x + y = 6$ Substitute $x = 1 - y$ in the second equation $-2(1 - y) + y = 5 \rightarrow y = 2$ Substitute $y = 2$ in $x = 1 + y$ $x = 1 + 2 = 3$	**Example:** $-2x - 2y = -13$ $-4x + 2y = 10$ (0.5, 6)

Solve each system of equation by substitution.

1) $-2x + 2y = 4$

$-2x + y = 3$

2) $-10x + 2y = -6$

$6x - 16y = 48$

3) $y = -8$

$16x - 12y = 72$

4) $2y = -6x + 10$

$10x - 8y = -6$

5) $3x - 9y = -3$

$3y = 3x - 3$

6) $-4x + 12y = 12$

$-14x + 16y = -10$

7) $x + 20y = 20$

$x + 15y = 5$

8) $2x + 8y = 28$

$x - 2y = 5$

Solving Systems of Equations by Elimination

Helpful Hints

- The elimination method for solving systems of linear equations uses the addition property of equality. You can add the same value to each side of an equation.

Example:

$$\begin{array}{r} \boldsymbol{x + 2y = 6} \\ \boldsymbol{+ \; -x + y = 3} \\ \hline 3y = 9 \\ y = 3 \\ \\ x + 6 = 6 \\ x = 0 \end{array}$$

Solve each system of equation by elimination.

1) $10x - 9y = -12$

$-5x + 3y = 6$

2) $-3x - 4y = 5$

$x - 2y = 5$

3) $5x - 14y = 22$

$-6x + 7y = 3$

4) $10x - 14y = -4$

$-10x - 20y = -30$

5) $32x + 14y = 52$

$16x - 4y = -40$

6) $2x - 8y = -6$

$8x + 2y = 10$

7) $-4x + 4y = -4$

$4x + 2y = 10$

8) $4x + 6y = 10$

$8x + 12y = -20$

Systems of Equations Word Problems

Helpful Hints

Define your variables, Write two equations, and use one of the methods for solving systems of equations to solve.

Example:

The difference of two numbers is 6. Their sum is 14. Find the numbers.

$x + y = 6$

$x + y = 14$ (10, 4)

Solve.

1) A farmhouse shelters 10 animals, some are pigs and some are ducks. Altogether there are 36 legs. How many of each animal are there?

2) A class of 195 students went on a field trip. They took vehicles, some cars and some buses. Find the number of cars and the number of buses they took if each car holds 5 students and each bus hold 45 students.

3) The difference of two numbers is 6. Their sum is 14. Find the numbers.

4) The sum of the digits of a certain two–digit number is 7. Reversing its increasing the number by 9. What is the number?

5) The difference of two numbers is 18. Their sum is 66. Find the numbers.

Answers of Worksheets – Chapter 11

Solving Systems of Equations by Substitution

1) $(-1, 1)$
2) $(0, -3)$
3) $(-4, -8)$
4) $(1, 2)$
5) $(2, 1)$
6) $(3, 2)$
7) $(-4, 3)$
8) $(8, \frac{3}{2})$

Solving Systems of Equations by Elimination

1) $(-1.2, 0)$
2) $(1, -2)$
3) $(-4, -3)$
4) $(1, 1)$
5) $(-1, 6)$
6) $(1, 1)$
7) $(2, 1)$
8) No solution

Systems of Equations Word Problems

1) There are 8 pigs and 2 ducks.
2) There are 3 cars and 4 buses.
3) 10 and 4.
4) 34
5) 24 and 42

Chapter 12: Exponents and Radicals

Topics that you'll learn in this chapter:

- ✓ Multiplication Property of Exponents
- ✓ Division Property of Exponents
- ✓ Powers of Products and Quotients
- ✓ Zero and Negative Exponents
- ✓ Negative Exponents and Negative Bases
- ✓ Writing Scientific Notation
- ✓ Square Roots

Mathematics is no more computation than typing is literature.

– John Allen Paulos

Multiplication Property of Exponents

Helpful Hints

Exponents rules

$x^a \cdot x^b = x^{a+b}$ $\frac{x^a}{x^b} = x^{a-b}$

$\frac{1}{x^b} = x^{-b}$ $(x^a)^b = x^{a.b}$

$(xy)^a = x^a \cdot y^a$

Example:

$(x^2y)^3 = x^6y^3$

Simplify.

1) $4^2 \cdot 4^2$

2) $2 \cdot 2^2 \cdot 2^2$

3) $3^2 \cdot 3^2$

4) $3x^3 \cdot x$

5) $12x^4 \cdot 3x$

6) $6x \cdot 2x^2$

7) $5x^4 \cdot 5x^4$

8) $6x^2 \cdot 6x^3y^4$

9) $7x^2y^5 \cdot 9xy^3$

10) $7xy^4 \cdot 4x^3y^3$

11) $(2x^2)^2$

12) $3x^5y^3 \cdot 8x^2y^3$

13) $7x^3 \cdot 10y^3x^5 \cdot 8yx^3$

14) $(x^4)^3$

15) $(2x^2)^4$

16) $(x^2)^3$

17) $(6x)^2$

18) $3x^4y^5 \cdot 7x^2y^3$

Division Property of Exponents

Helpful Hints

$\frac{x^a}{x^b} = x^{a-b}$, $x \neq 0$

Example:

$\frac{x^{12}}{x^5} = x^7$

Simplify.

1) $\frac{5^5}{5}$

2) $\frac{3}{3^5}$

3) $\frac{2^2}{2^3}$

4) $\frac{2^4}{2^2}$

5) $\frac{x}{x^3}$

6) $\frac{3x^3}{9x^4}$

7) $\frac{2x^{-5}}{9x^{-2}}$

8) $\frac{21x^8}{7x^3}$

9) $\frac{7x^6}{4x^7}$

10) $\frac{6x^2}{4x^3}$

11) $\frac{5x}{10x^3}$

12) $\frac{3x^3}{2x^5}$

13) $\frac{12x^3}{14x^6}$

14) $\frac{12x^3}{9y^8}$

15) $\frac{25xy^4}{5x^6y^2}$

16) $\frac{2x^4}{7x}$

17) $\frac{16x^2y^8}{4x^3}$

18) $\frac{12x^4}{15x^7y^9}$

19) $\frac{12yx^4}{10yx^8}$

20) $\frac{16x^4y}{9x^8y^2}$

21) $\frac{5x^8}{20x^8}$

Powers of Products and Quotients

Helpful Hints

For any nonzero numbers a and b and any integer $(ab)^x = a^x b^x$

Example:

$(2x^2 \cdot y^3)^2 =$

$4x^2 \cdot y^6$

Simplify.

1) $(2x^3)^4$

2) $(4xy^4)^2$

3) $(5x^4)^2$

4) $(11x^5)^2$

5) $(4x^2y^4)^4$

6) $(2x^4y^4)^3$

7) $(3x^2y^2)^2$

8) $(3x^4y^3)^4$

9) $(2x^6y^8)^2$

10) $(12x\ 3x)^3$

11) $(2x^9\ x^6)^3$

12) $(5x^{10}y^3)^3$

13) $(4x^3\ x^2)^2$

14) $(3x^3\ 5x)^2$

15) $(10x^{11}y^3)^2$

16) $(9x^7\ y^5)^2$

17) $(4x^4y^6)^5$

18) $(4x^4)^2$

19) $(3x\ 4y^3)^2$

20) $(9x^2y)^3$

21) $(12x^2y^5)^2$

Zero and Negative Exponents

Helpful Hints

A negative exponent simply means that the base is on the wrong side of the fraction line, so you need to flip the base to the other side. For instance, "x^{-2}" (pronounced as "ecks to the minus two") just means "x^2" but underneath, as in $\frac{1}{x^2}$

Example:

$5^{-2} = \frac{1}{25}$

Evaluate the following expressions.

1) 8^{-2}

2) 2^{-4}

3) 10^{-2}

4) 5^{-3}

5) 22^{-1}

6) 9^{-1}

7) 3^{-2}

8) 4^{-2}

9) 5^{-2}

10) 35^{-1}

11) 6^{-3}

12) 0^{15}

13) 10^{-9}

14) 3^{-4}

15) 5^{-2}

16) 2^{-3}

17) 3^{-3}

18) 8^{-1}

19) 7^{-3}

20) 6^{-2}

21) $(\frac{2}{3})^{-2}$

22) $(\frac{1}{5})^{-3}$

23) $(\frac{1}{2})^{-8}$

24) $(\frac{2}{5})^{-3}$

25) 10^{-3}

26) 1^{-10}

Negative Exponents and Negative Bases

Helpful Hints

– Make the power positive. A negative exponent is the reciprocal of that number with a positive exponent.

– The parenthesis is important!

-5^{-2} is not the same as $(-5)^{-2}$

$-5^{-2} = -\frac{1}{5^2}$ and $(-5)^{-2} = +\frac{1}{5^2}$

Example:

$2x^{-3} = \frac{2}{x^3}$

Simplify.

1) -6^{-1}

2) $-4x^{-3}$

3) $-\frac{5x}{x^{-3}}$

4) $-\frac{a^{-3}}{b^{-2}}$

5) $-\frac{5}{x^{-3}}$

6) $\frac{7b}{-9c^{-4}}$

7) $-\frac{5n^{-2}}{10p^{-3}}$

8) $\frac{4ab^{-2}}{-3c^{-2}}$

9) $-12x^2y^{-3}$

10) $(-\frac{1}{3})^{-2}$

11) $(-\frac{3}{4})^{-2}$

12) $(\frac{3a}{2c})^{-2}$

13) $(-\frac{5x}{3yz})^{-3}$

14) $-\frac{2x}{a^{-4}}$

Writing Scientific Notation

Helpful Hints

– It is used to write very big or very small numbers in decimal form.

– In scientific notation all numbers are written in the form of:

$$m \times 10^n$$

Decimal notation	Scientific notation
5	5×10^0
–25,000	-2.5×10^4
0.5	5×10^{-1}
2,122.456	$2{,}122456 \times 10^3$

Write each number in scientific notation.

1) 91×10^3
2) 60
3) 2000000
4) 0.0000006
5) 354000
6) 0.000325
7) 2.5
8) 0.00023
9) 56000000
10) 2000000
11) 78000000
12) 0.0000022
13) 0.00012
14) 0.004
15) 78
16) 1600
17) 1450
18) 130000
19) 60
20) 0.113
21) 0.02

Square Roots

Helpful Hints		
	– A square root of x is a number r whose square is: $r^2 = x$	**Example:**
	r is a square root of x.	$\sqrt{4} = 2$

Find the value each square root.

1) $\sqrt{1}$

2) $\sqrt{4}$

3) $\sqrt{9}$

4) $\sqrt{25}$

5) $\sqrt{16}$

6) $\sqrt{49}$

7) $\sqrt{36}$

8) $\sqrt{0}$

9) $\sqrt{64}$

10) $\sqrt{81}$

11) $\sqrt{121}$

12) $\sqrt{225}$

13) $\sqrt{144}$

14) $\sqrt{100}$

15) $\sqrt{256}$

16) $\sqrt{289}$

17) $\sqrt{324}$

18) $\sqrt{400}$

19) $\sqrt{900}$

20) $\sqrt{529}$

21) $\sqrt{90}$

Answers of Worksheets – Chapter 12

Multiplication Property of Exponents

1) 4^4
2) 2^5
3) 3^4
4) $3x^4$
5) $36x^5$
6) $12x^3$
7) $25x^8$
8) $36x^5y^4$
9) $63x^3y^8$
10) $28x^4y^7$
11) $4x^4$
12) $24x^7y^6$
13) $560x^{11}y^4$
14) x^{12}
15) $16x^8$
16) x^6
17) $36x^2$
18) $21x^6y^8$

Division Property of Exponents

1) 5^4
2) $\frac{1}{3^4}$
3) $\frac{1}{2}$
4) 2^2
5) $\frac{1}{x^2}$
6) $\frac{1}{3x}$
7) $\frac{2}{9x^3}$
8) $3x^5$
9) $\frac{7}{4x}$
10) $\frac{3}{2x}$
11) $\frac{1}{2x^2}$
12) $\frac{3}{2x^2}$
13) $\frac{6}{7x^3}$
14) $\frac{4x^3}{3y^8}$
15) $\frac{5y^2}{x^5}$
16) $\frac{2x^3}{7}$
17) $\frac{4y^8}{x}$
18) $\frac{4}{5x^3y^9}$
19) $\frac{6}{5x^4}$
20) $\frac{16}{9x^4y}$
21) $\frac{1}{4}$

Powers of Products and Quotients

1) $16x^{12}$
2) $16x^2y^8$
3) $25x^8$
4) $121x^{10}$
5) $256x^8y^{16}$
6) $8x^{12}y^{12}$
7) $9x^4y^4$
8) $81x^{16}y^{12}$
9) $4x^{12}y^{16}$
10) $46{,}656x^6$
11) $8x^{45}$
12) $125x^{30}y^9$
13) $16x^{10}$
14) $225x^8$
15) $100x^{22}y^6$
16) $81x^{14}y^{10}$
17) $1{,}024x^{20}y^{30}$
18) $16x^8$
19) $144x^2y^6$
20) $729x^6y^3$
21) $144x^4y^{10}$

Zero and Negative Exponents

1) $\frac{1}{64}$
2) $\frac{1}{16}$
3) $\frac{1}{100}$
4) $\frac{1}{125}$
5) $\frac{1}{22}$
6) $\frac{1}{9}$
7) $\frac{1}{9}$
8) $\frac{1}{16}$
9) $\frac{1}{25}$
10) $\frac{1}{35}$
11) $\frac{1}{216}$
12) 0
13) $\frac{1}{1000000000}$
14) $\frac{1}{81}$
15) $\frac{1}{25}$
16) $\frac{1}{8}$
17) $\frac{1}{27}$
18) $\frac{1}{8}$
19) $\frac{1}{343}$
20) $\frac{1}{36}$
21) $\frac{9}{4}$
22) 125
23) 256
24) $\frac{125}{8}$
25) $\frac{1}{1000}$
26) 1

Negative Exponents and Negative Bases

1) $-\frac{1}{6}$
2) $-\frac{4}{x^3}$
3) $-5x^4$
4) $-\frac{b^2}{a^3}$
5) $-5x^3$
6) $-\frac{7bc^4}{9}$
7) $-\frac{p^3}{2n^2}$
8) $-\frac{4ac^2}{3b^2}$
9) $-\frac{12x^2}{y^3}$
10) 9
11) $\frac{16}{9}$
12) $\frac{4c^2}{9a^2}$
13) $-\frac{27y^3z^3}{125x^3}$
14) $-2xa^4$

Writing Scientific Notation

1) 9.1×10^4
2) 6×10^1
3) 2×10^6
4) 6×10^{-7}
5) 3.54×10^5
6) 3.25×10^{-4}
7) 2.5×10^0
8) 2.3×10^{-4}
9) 5.6×10^7
10) 2×10^6
11) 7.8×10^7
12) 2.2×10^{-6}
13) 1.2×10^{-4}
14) 4×10^{-3}
15) 7.8×10^1

16) 1.6×10^3

17) 1.45×10^3

18) 1.3×10^5

19) 6×10^1

20) 1.13×10^{-1}

21) 2×10^{-2}

Square Roots

1) 1
2) 2
3) 3
4) 5
5) 4
6) 7
7) 6
8) 0
9) 8
10) 9
11) 11
12) 15
13) 12
14) 10
15) 16
16) 17
17) 18
18) 20
19) 30
20) 23
21) $3\sqrt{10}$

Chapter 13: Geometry

Topics that you'll learn in this chapter:

- ✓ The Pythagorean Theorem
- ✓ Area of Triangles
- ✓ Perimeter of Polygons
- ✓ Area and Circumference of Circles
- ✓ Area of Squares, Rectangles, and Parallelograms
- ✓ Area of Trapezoids

Mathematics is, as it were, a sensuous logic, and relates to philosophy as do the arts, music, and plastic art to poetry. — K. Shegel

The Pythagorean Theorem

Helpful Hints

– In any right triangle:

$a^2 + b^2 = c^2$

Example:

Missing side = 6

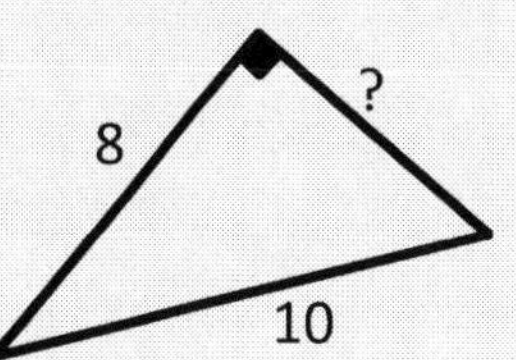

Do the following lengths form a right triangle?

1)

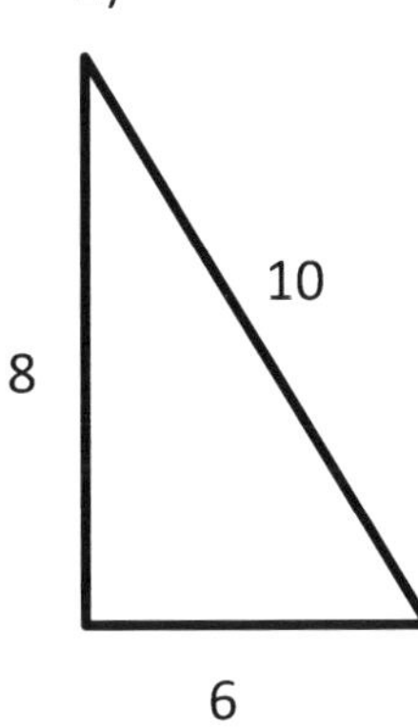

2)

3)

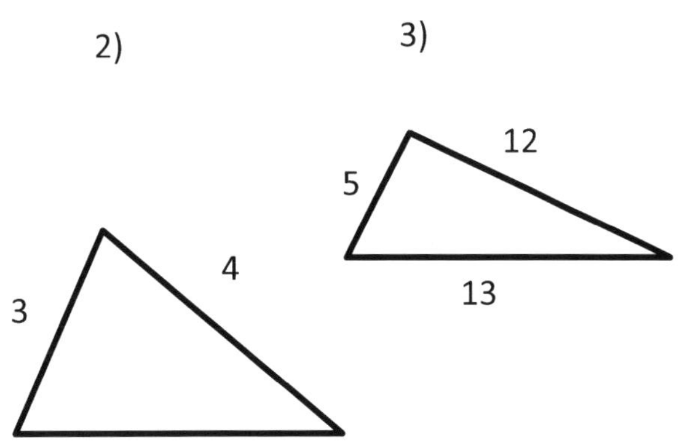

Find each missing length to the nearest tenth.

4)

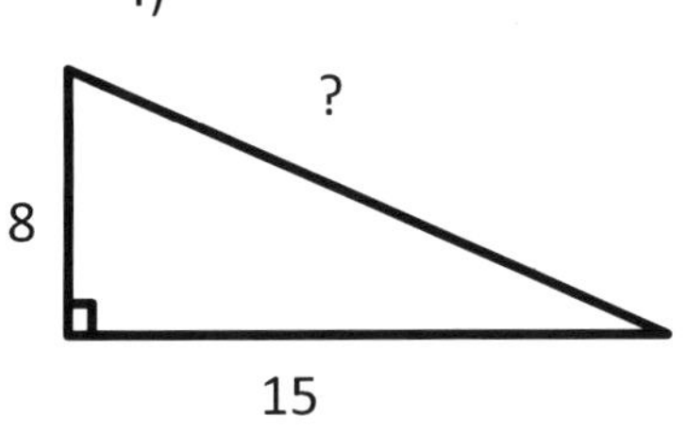

5)

6)

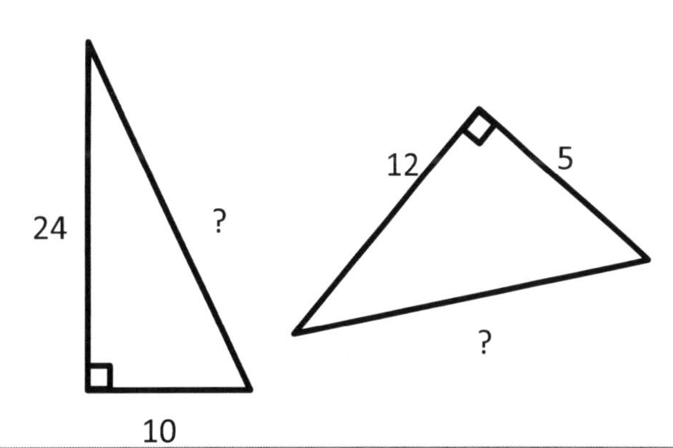

Area of Triangles

Helpful Hints

Area $= \frac{1}{2}$ $(base \times height)$

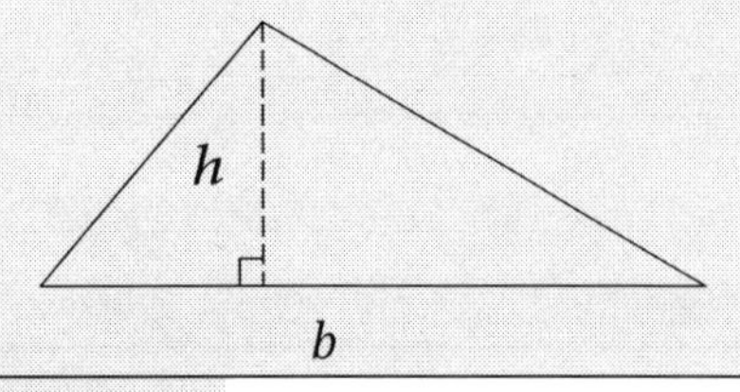

Find the area of each.

1)

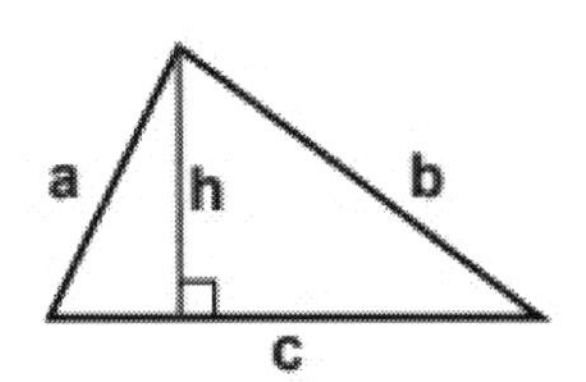

c = 9 mi

h = 3.7 mi

2)

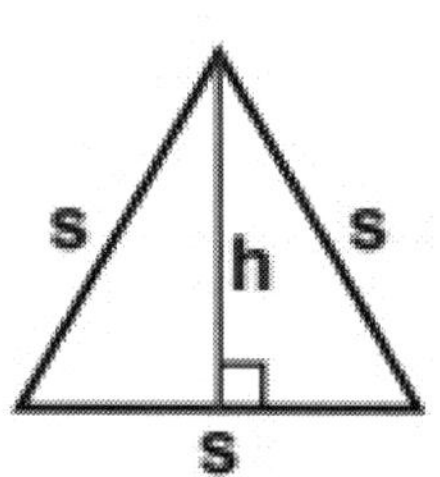

s = 14 m

h = 12.2 m

3)

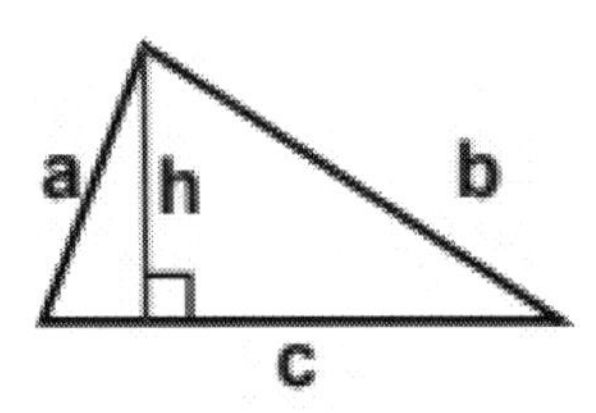

a = 5 m

b = 11 m

c = 14 m

h = 4 m

4)

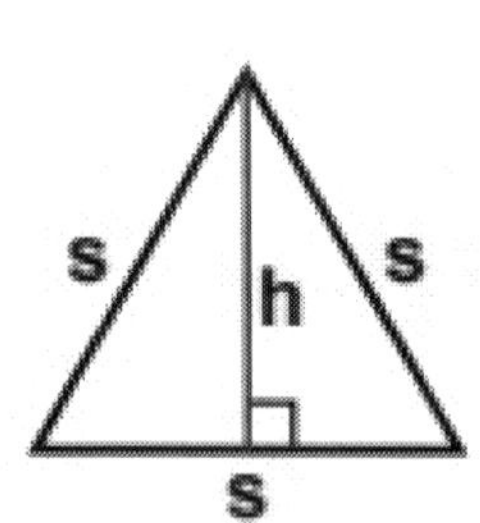

s = 10 m

h = 8.6 m

Perimeter of Polygons

Helpful Hints

Perimeter of a square = 4s

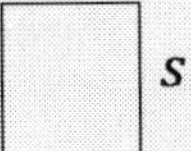

Perimeter of a rectangle

$= 2(l + w)$

w

l

Perimeter of trapezoid

= a + b + c + d

a

d b

c

Perimeter of Pentagon = 6a

a

Perimeter of a parallelogram = 2(l + w)

l

w

Example:

P = 18

3 m

3 m 3 m

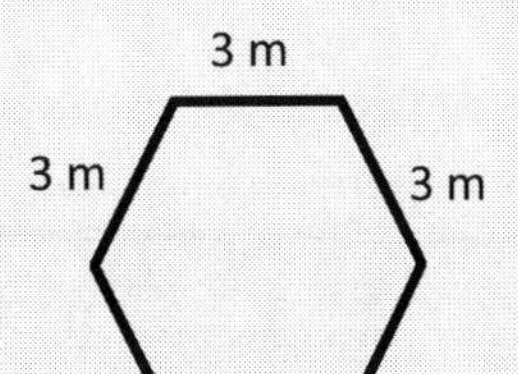

Find the perimeter of each shape.

1)

5 m

5 m 5 m

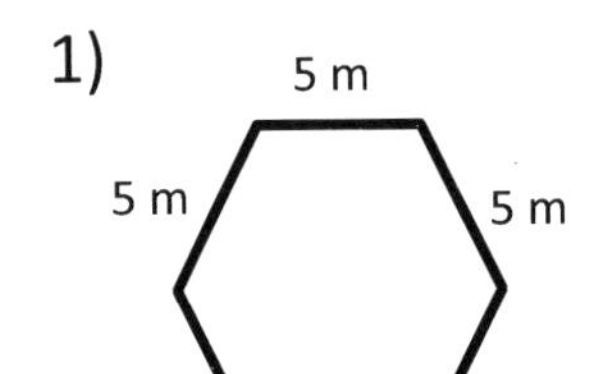

2)

15 mm

15 mm 15mm

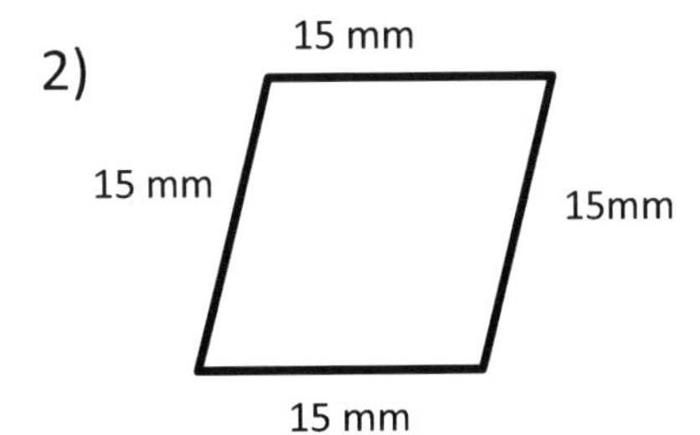

15 mm

3)

12 ft 12 ft

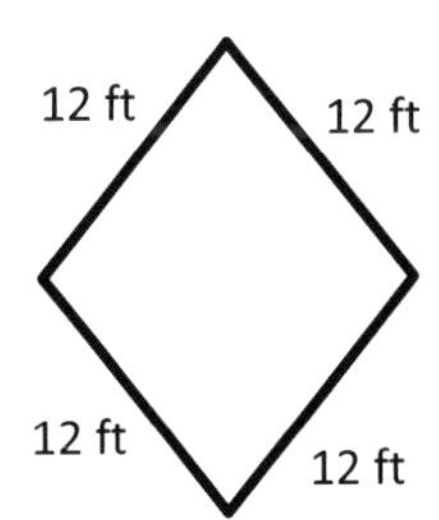

12 ft 12 ft

4)

18 in

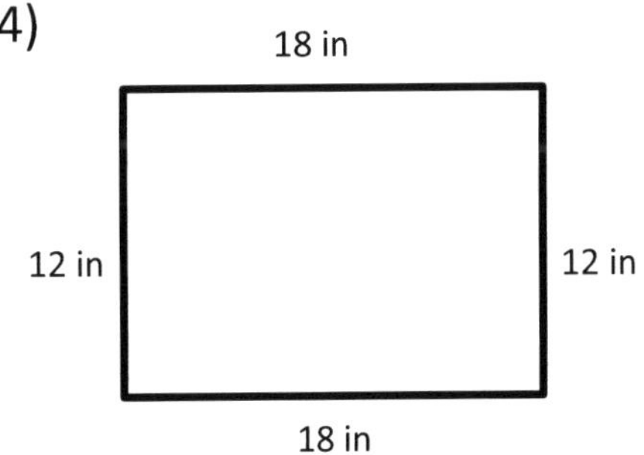

12 in 12 in

18 in

Area and Circumference of Circles

Helpful Hints

Area = πr^2

Circumference = $2\pi r$

r

Example:

If the radius of a circle is 3, then:

Area = 28.27

Circumference = 18.85

Find the area and circumference of each. ($\pi = 3.14$)

1)

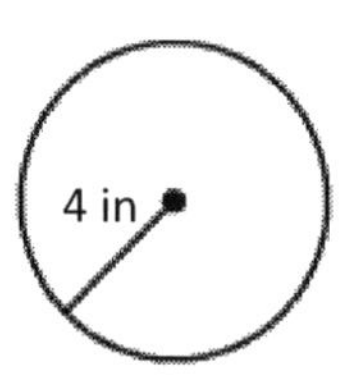

2)

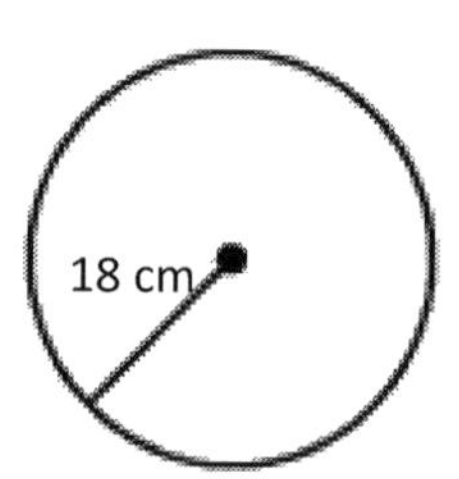

3)

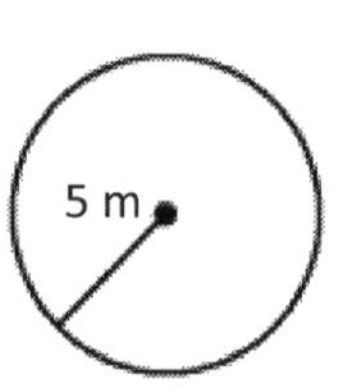

4)

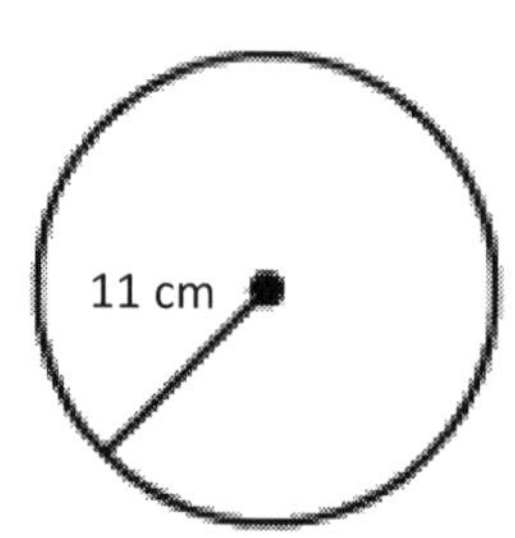

5)

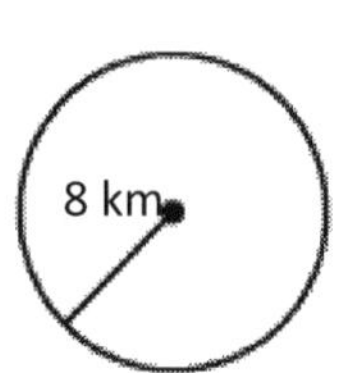

6)

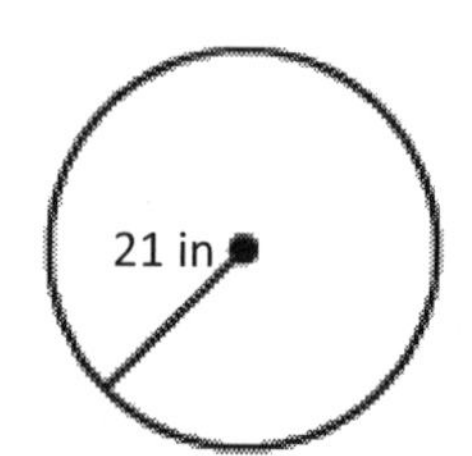

Area of Squares, Rectangles, and Parallelograms

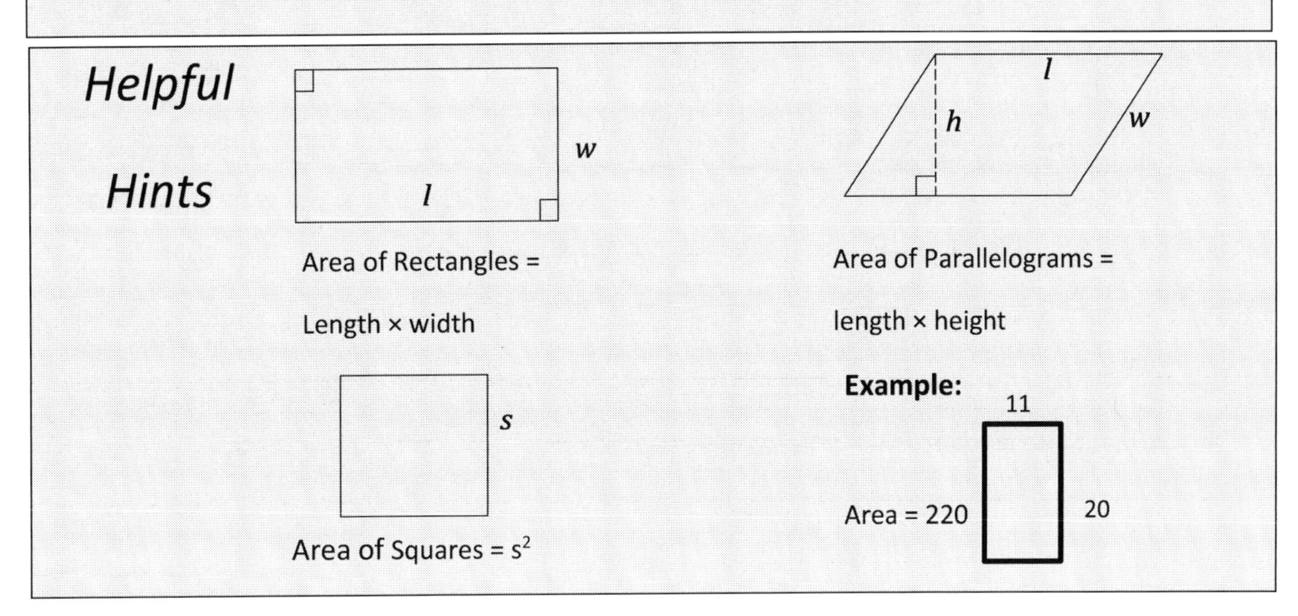

Find the area of each.

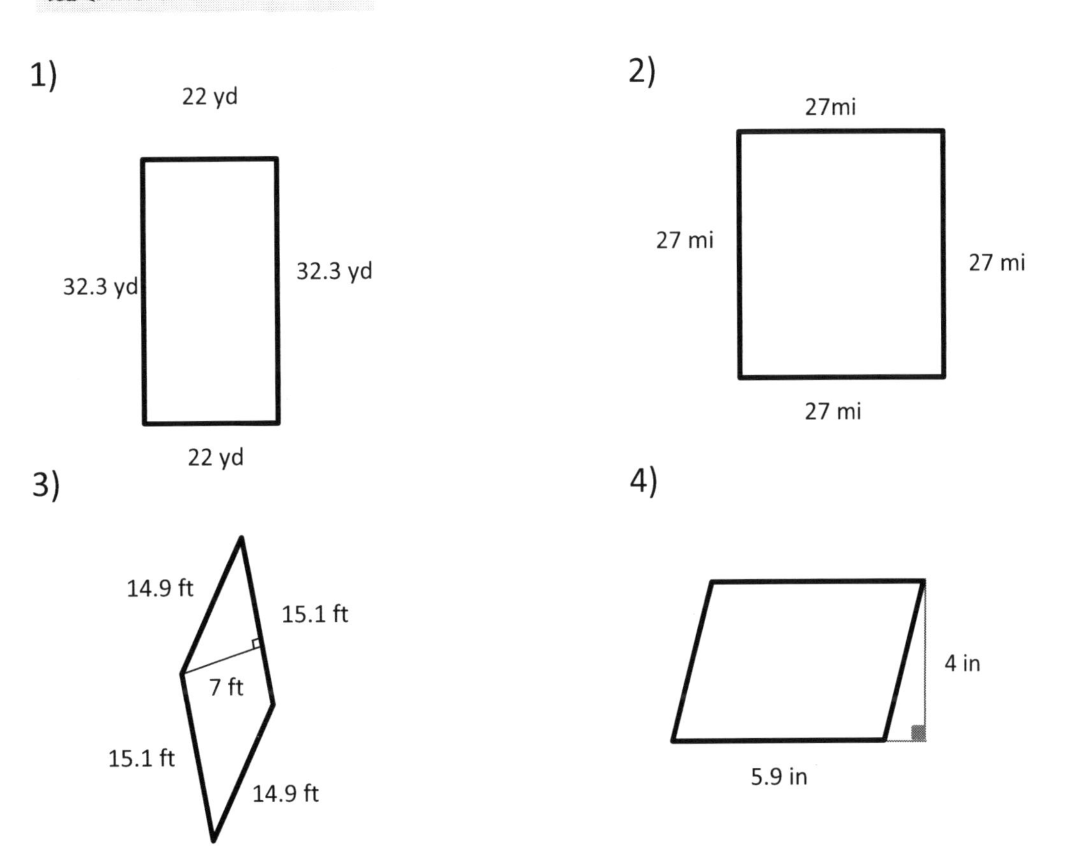

Area of Trapezoids

Helpful Hints

$$A = \frac{1}{2}h(b_1 + b_2)$$

Example:

16 cm

18 cm

A = 252 cm^2

12 cm

Calculate the area for each trapezoid.

1)

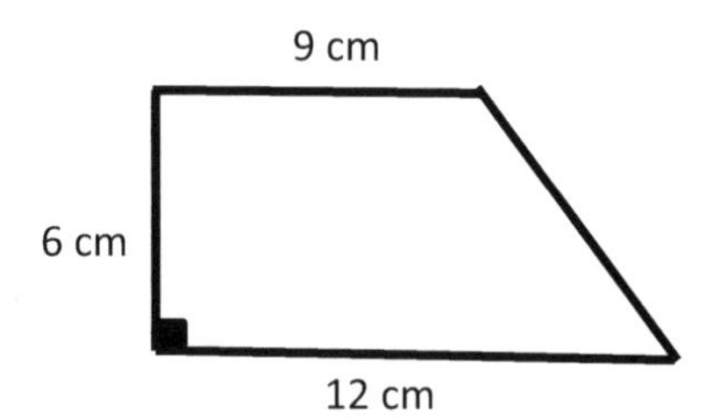

2)

14 m

10 m

18 m

3)

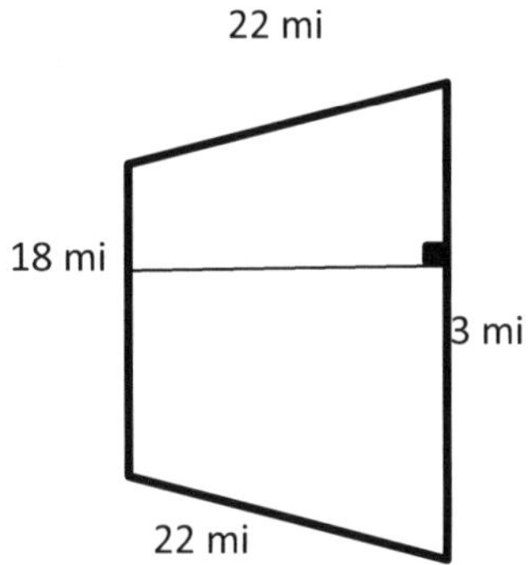

4)

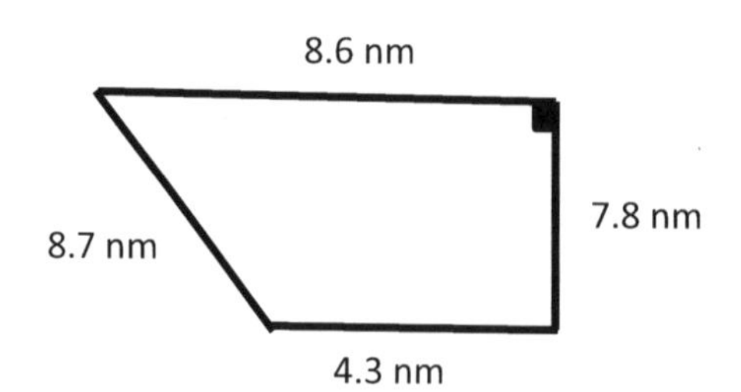

Answers of Worksheets – Chapter 13

The Pythagorean Theorem

1) yes
2) yes
3) yes
4) 17
5) 26
6) 13

Area of Triangles

1) 16.65 mi^2
2) 85.4 m^2
3) 28 m^2
4) 43 m^2

Perimeter of Polygons

1) 30 m
2) 60 mm
3) 48 ft
4) 60 in

Area and Circumference of Circles

1) Area: 50.24 in^2, Circumference: 25.12 in
2) Area: 1,017.36 cm^2, Circumference: 113.04 cm
3) Area: 78.5m^2, Circumference: 31.4 m
4) Area: 379.94 cm^2, Circumference: 69.08 cm
5) Area: 200.96 km^2, Circumference: 50.2 km
6) Area: 1,384.74 km^2, Circumference: 131.88 km

Area of Squares, Rectangles, and Parallelograms

1) 710.6 yd^2
2) 729 mi^2
3) 105.7 ft^2
4) 23.6 in^2

Area of Trapezoids

1) 63 cm^2
2) 160 m^2
3) 410 mi^2
4) 50.31 nm^2

Chapter 14: Solid Figures

Topics that you'll learn in this chapter:

- ✓ Volume of Cubes
- ✓ Volume of Rectangle Prisms
- ✓ Surface Area of Cubes
- ✓ Surface Area of Rectangle Prisms
- ✓ Volume of a Cylinder
- ✓ Surface Area of a Cylinder

Mathematics is a great motivator for all humans. Because its career starts with zero and it never end (infinity)

Volume of Cubes

Helpful Hints	
	– Volume is the measure of the amount of space inside of a solid figure, like a cube, ball, cylinder or pyramid.
	– Volume of a cube = $(\text{one side})^3$
	– Volume of a rectangle prism: Length × Width × Height

Find the volume of each.

1) 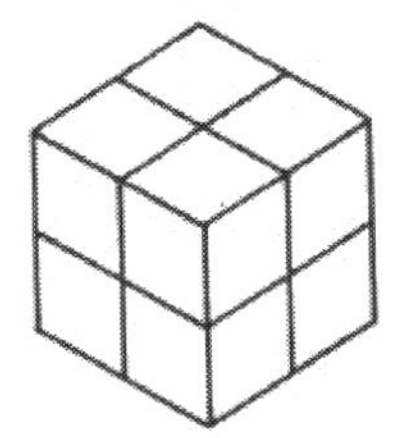

2) 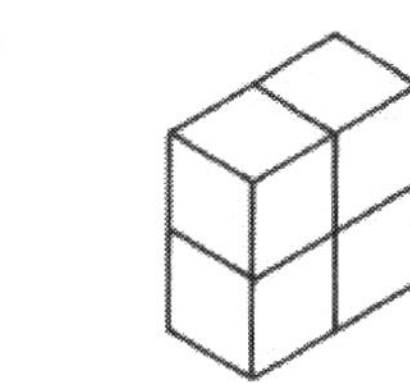

3) 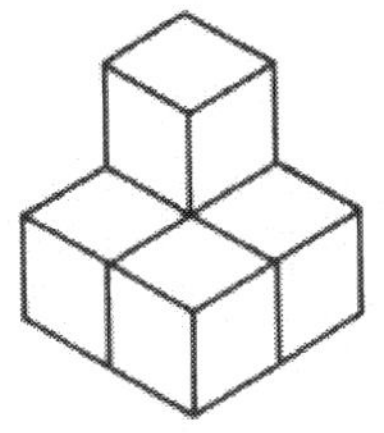

4) 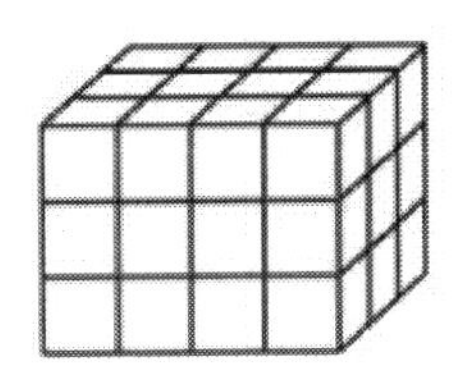

5) 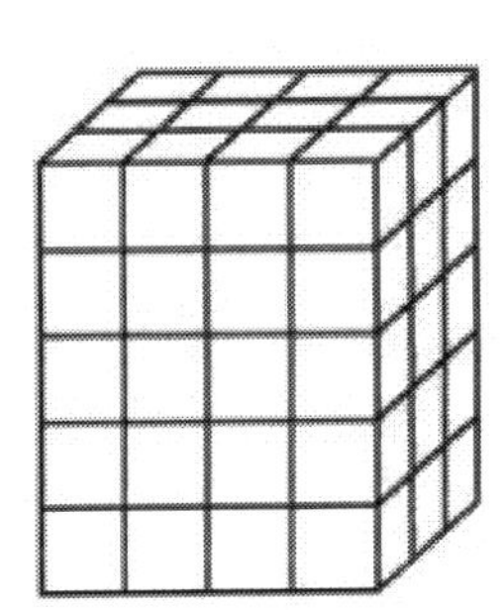

6) 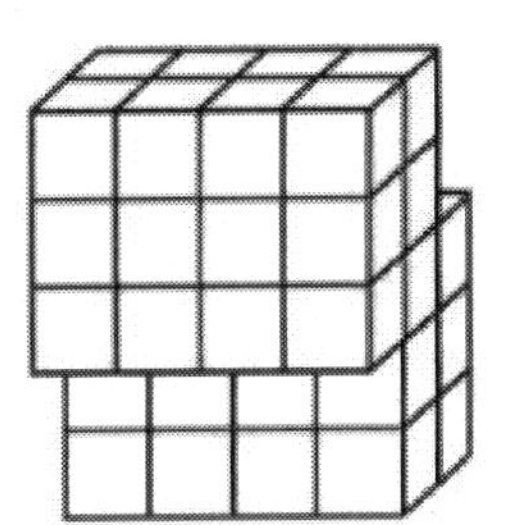

Volume of Rectangle Prisms

Helpful Hints

Volume of rectangle prism

length × width × height

Example:

$10 \times 5 \times 8 = 400m^3$

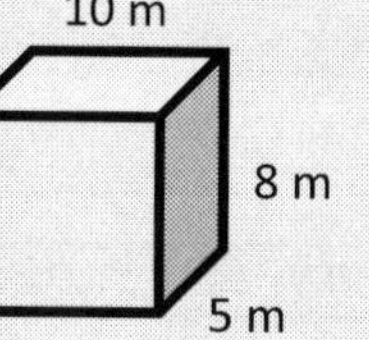

Find the volume of each of the rectangular prisms.

1)

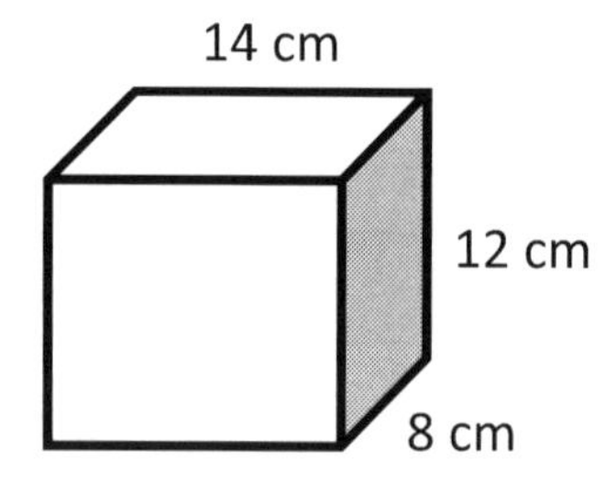

2)

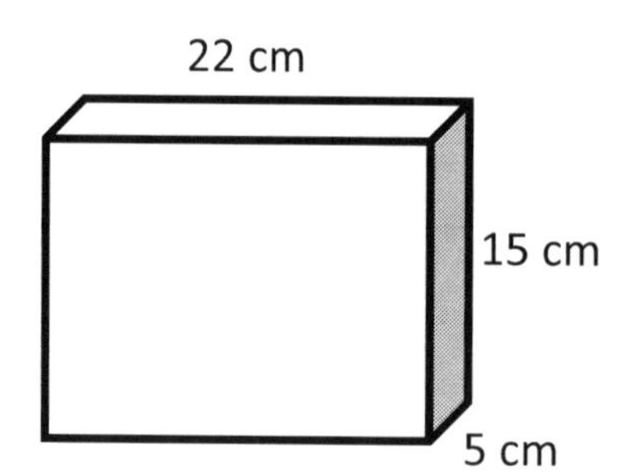

3)

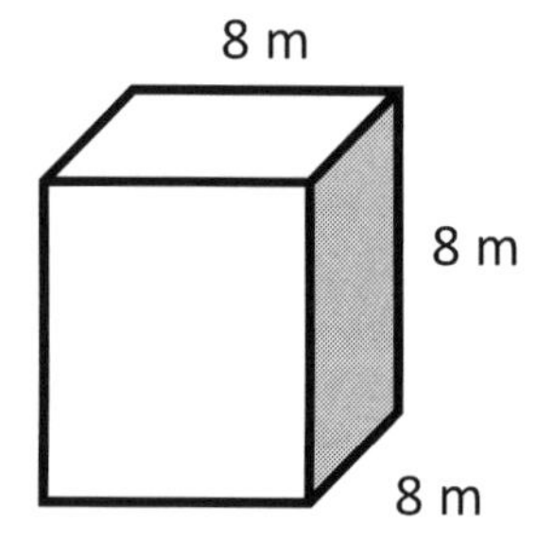

4)

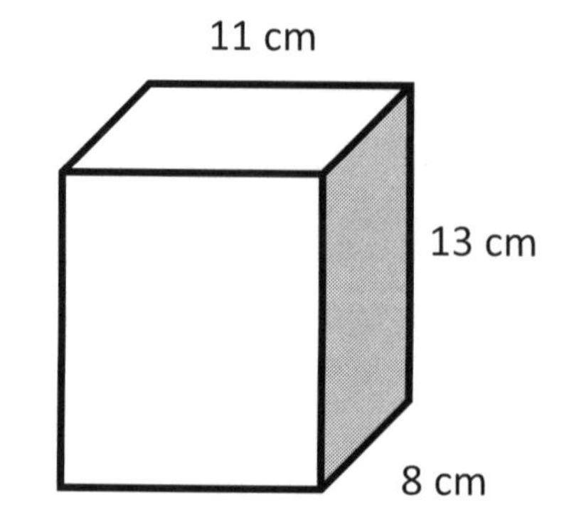

Surface Area of Cubes

Helpful Hints

Surface Area of a cube =

6 × (one side of the cube)^2

Example:

$6 \times 4^2 = 96m^2$

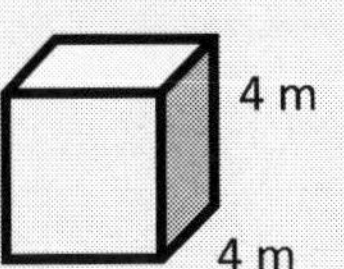

Find the surface of each cube.

1)

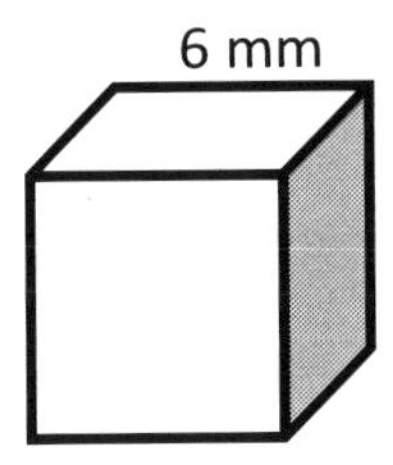

2)

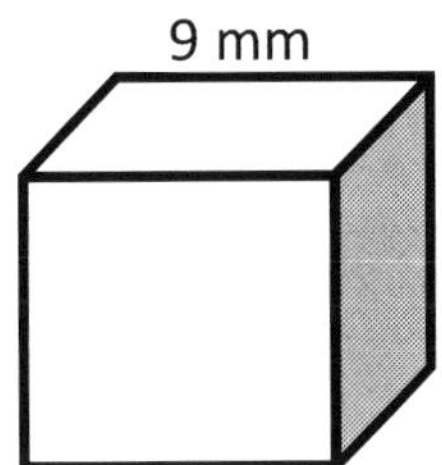

3)

4)

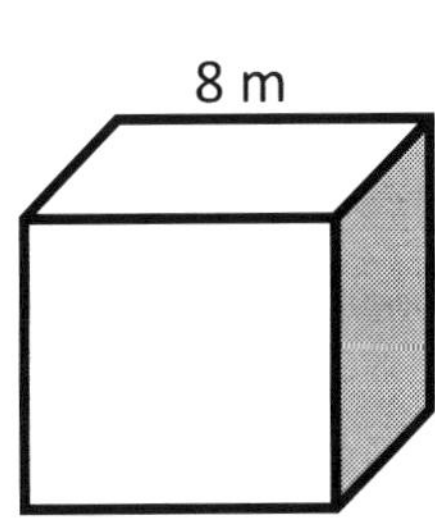

5)

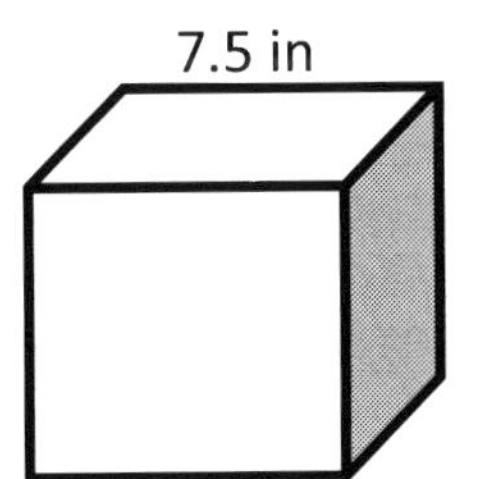

6)

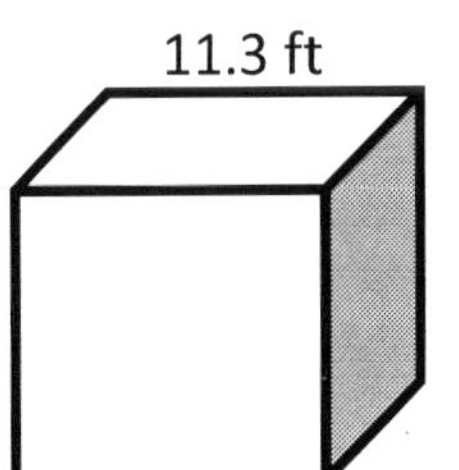

Surface Area of a Rectangle Prism

Helpful Hints

Surface Area of a Rectangle Prism Formula:

SA =2 [(width × length) + (height × length) + width × height)]

Find the surface of each prism.

1)

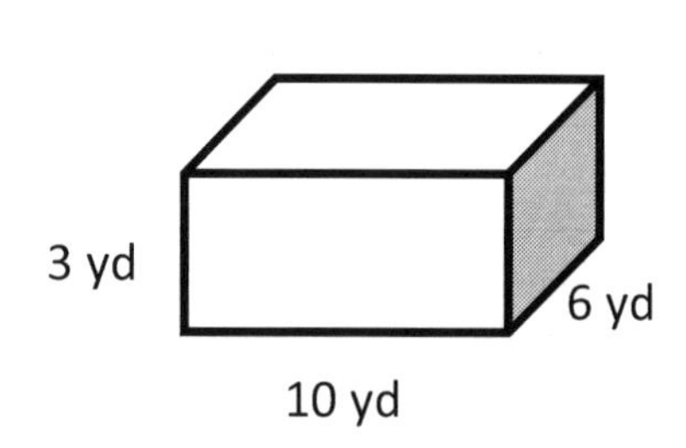

2)

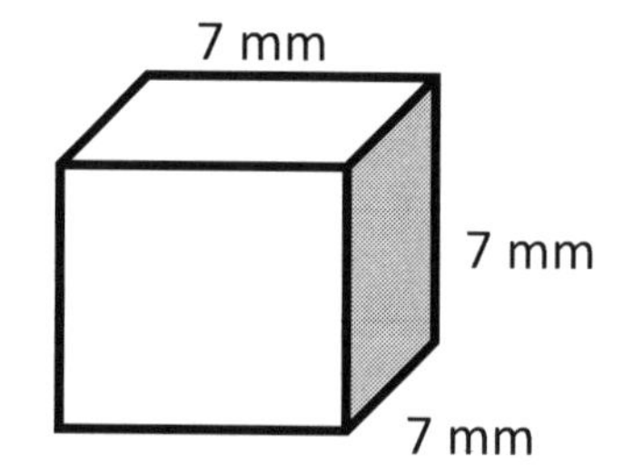

3)

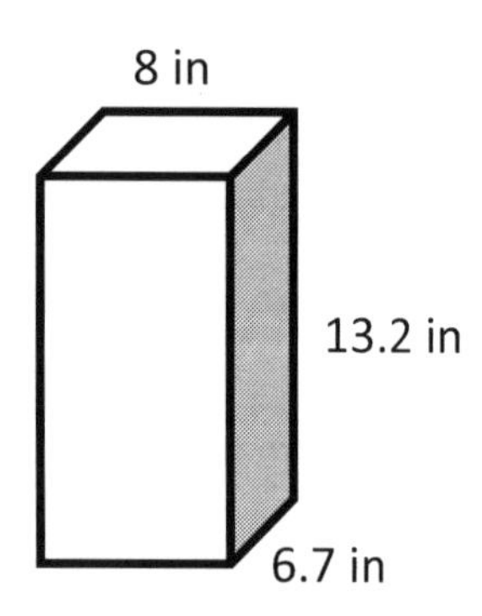

4)

17 cm

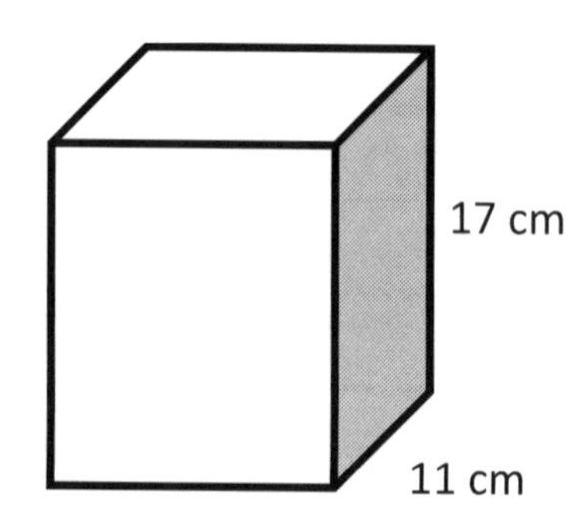

Volume of a Cylinder

Helpful Hints

Volume of Cylinder Formula = $\pi(\text{radius})^2 \times \text{height}$

$\pi = 3.14$

Find the volume of each cylinder. ($\pi = 3.14$)

1)

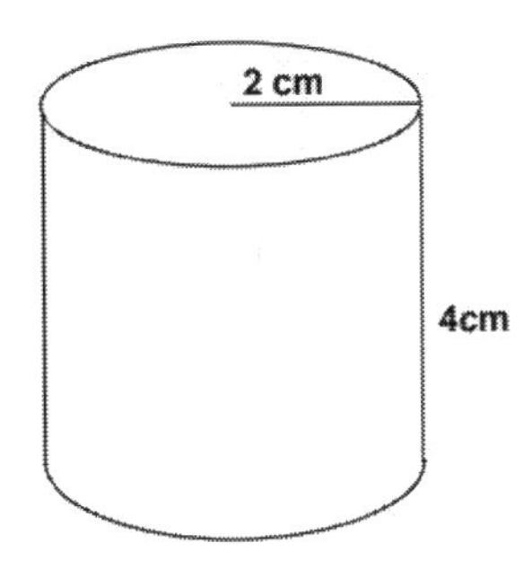

2)

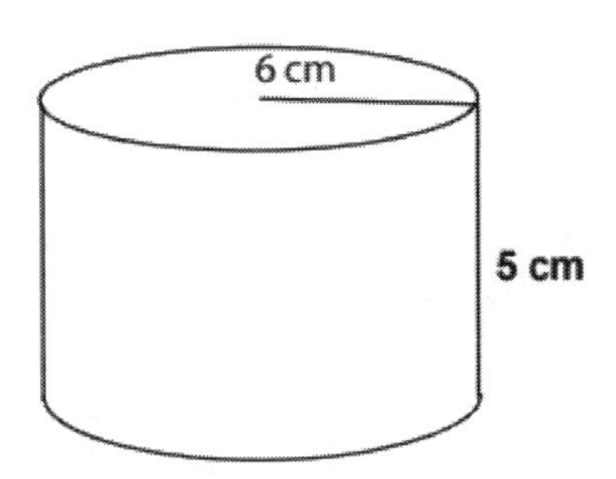

3)

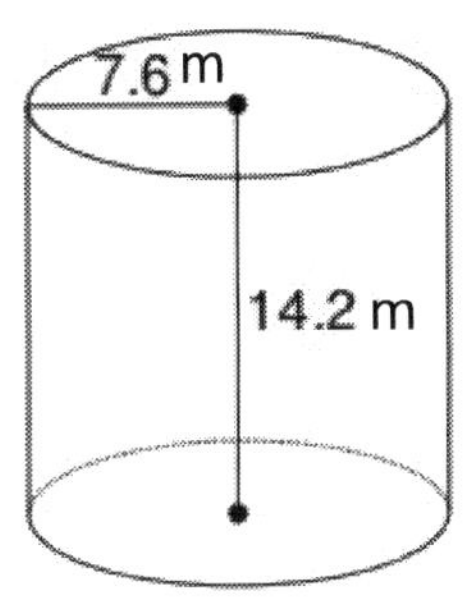

4)

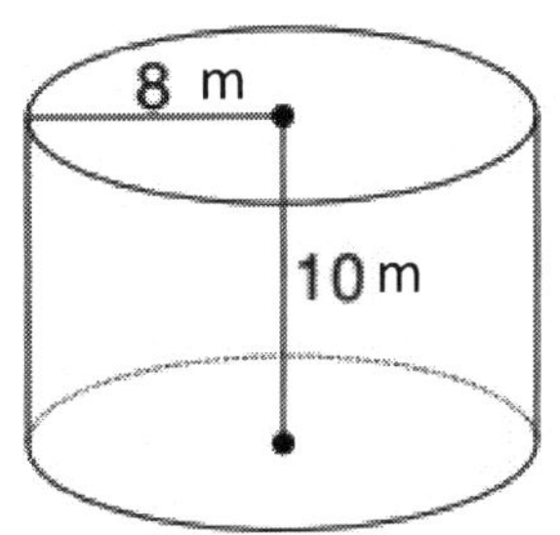

Surface Area of a Cylinder

Helpful Hints

Surface area of a cylinder

$SA = 2\pi r^2 + 2\pi rh$

Example:

Surface area

= 1727

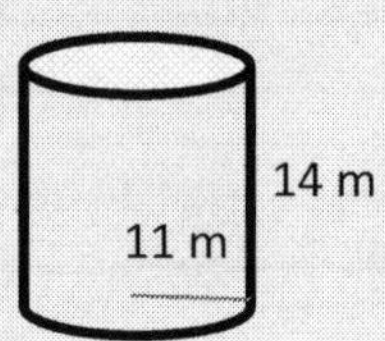

Find the surface of each cylinder. ($\pi = 3.14$)

1)

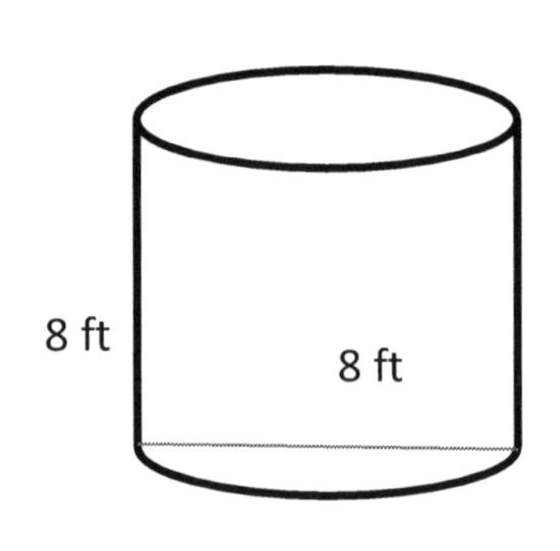

2)

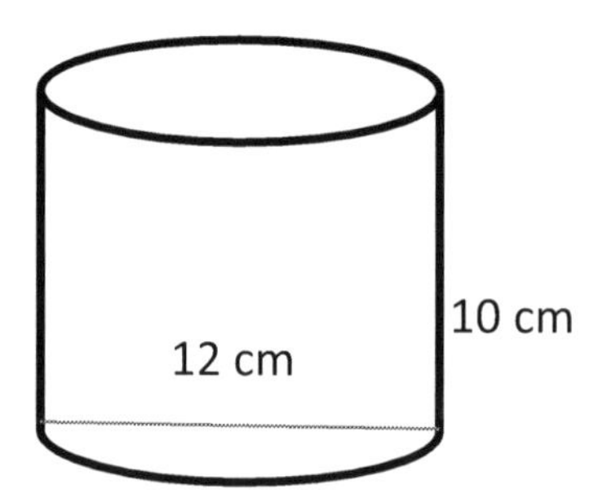

3)

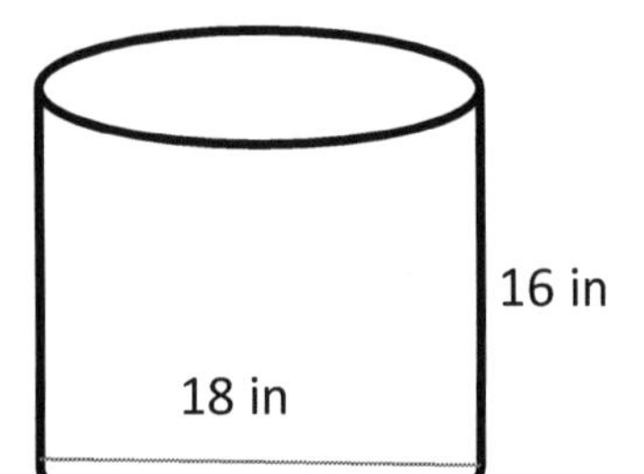

4)

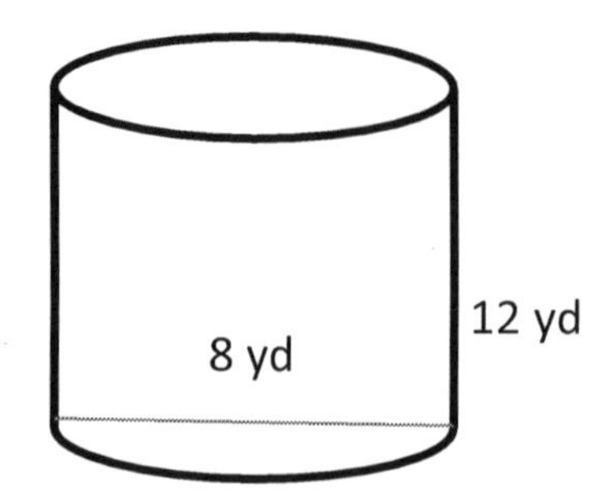

Answers of Worksheets – Chapter 14

Volumes of Cubes

1) 8
2) 4
3) 5
4) 36
5) 60
6) 44

Volume of Rectangle Prisms

1) 1344 cm^3
2) 1650 cm^3
3) 512 m^3
4) 1144 cm^3

Surface Area of a Cube

1) 216 mm^2
2) 486 mm^2
3) 600 cm^2
4) 384 m^2
5) 337.5 in^2
6) 766.14 ft^2

Surface Area of a Prism

1) 216 yd^2
2) 294 mm^2
3) 495.28 in^2
4) 1326 cm^2

Volume of a Cylinder

1) 50.24 cm^3
2) 565.2 cm^3
3) 2,575.403 m^3
4) 2009.6 m^3

Surface Area of a Cylinder

1) 301.44 ft^2
2) 602.88 cm^2
3) 1413 in^2
4) 401.92 yd^2

Chapter 15: Statistics

Topics that you'll learn in this chapter:

- ✓ Mean, Median, Mode, and Range of the Given Data
- ✓ Box and Whisker Plots
- ✓ Bar Graph
- ✓ Stem– And– Leaf Plot
- ✓ The Pie Graph or Circle Graph
- ✓ Scatter Plots
- ✓ Probability

Mathematics is no more computation than typing is literature.

– John Allen Paulos

Mean, Median, Mode, and Range of the Given Data

Helpful Hints

- Mean: $\frac{\text{sum of the data}}{\text{of data entires}}$
- Mode: value in the list that appears most often
- Range: largest value – smallest value

Example:

22, 16, 12, 9, 7, 6, 4, 6

Mean = 10.25

Mod = 6

Range = 18

Find Mean, Median, Mode, and Range of the Given Data.

1) 7, 2, 5, 1, 1, 2

2) 2, 2, 2, 3, 6, 3, 7, 4

3) 9, 4, 3, 1, 7, 9, 4, 6, 4

4) 8, 4, 2, 4, 3, 2, 4, 5

5) 8, 5, 7, 5, 7, 9, 8

6) 5, 1, 4, 4, 9, 2, 9, 2, 5, 1

7) 4, 1, 5, 9, 7, 7, 5, 4, 3, 5

8) 7, 5, 4, 9, 6, 7, 7, 5, 2

9) 2, 5, 5, 6, 2, 4, 7, 6, 4, 9

10) 10, 5, 2, 5, 4, 5, 8, 10

11) 5, 1, 5, 2, 2

12) 2, 3, 5, 9, 6

Box and Whisker Plots

Helpful Hints	Box–and–whisker plots display data including quartiles. - IQR – interquartile range shows the difference from Q1 to Q3. - Extreme Values are the smallest and largest values in a data set.

Example:

73, 84, 86, 95, 68, 67, 100, 94, 77, 80, 62, 79

Maximum: 100, Minimum: 62, Q_1: 70.5, Q_2: 79.5, Q_3: 90

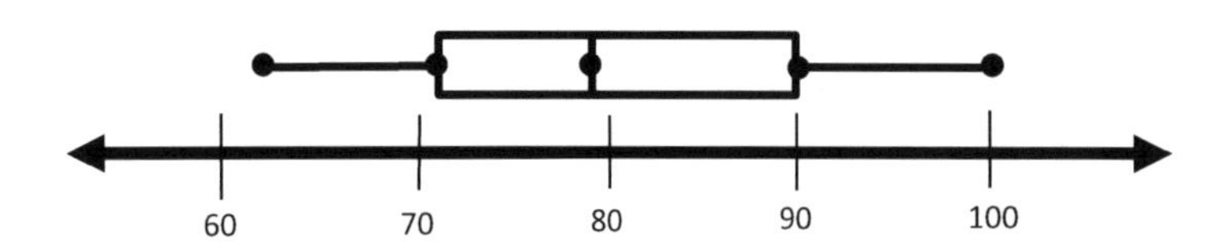

Make box and whisker plots for the given data.

11, 17, 22, 18, 23, 2, 3, 16, 21, 7, 8, 15, 5

Bar Graph

Helpful Hints

– A bar graph is a chart that presents data with bars in different heights to match with the values of the data. The bars can be graphed horizontally or vertically.

Graph the given information as a bar graph.

Day	Hot dogs sold
Monday	90
Tuesday	70
Wednesday	30
Thursday	20
Friday	60

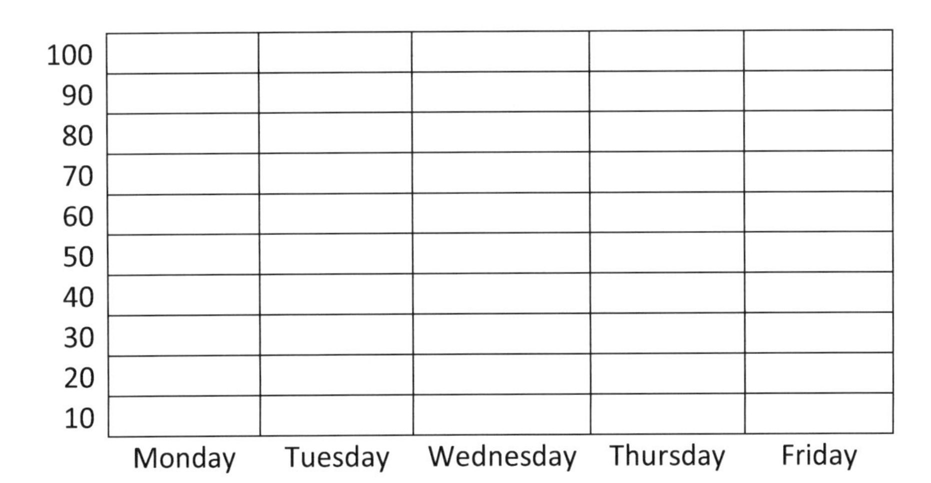

Stem–And–Leaf Plot

Helpful Hints

– Stem–and–leaf plots display the frequency of the values in a data set.

– We can make a frequency distribution table for the values, or we can use a stem–and–leaf plot.

Example:

56, 58, 42, 48, 66, 64, 53, 69, 45, 72

Stem	leaf
4	2 5 8
5	3 6 8
6	4 6 9
7	2

Make stem ad leaf plots for the given data.

1) 74, 88, 97, 72, 79, 86, 95, 79, 83, 91

Stem | Leaf plot

2) 37, 48, 26, 33, 49, 26, 19, 26, 48

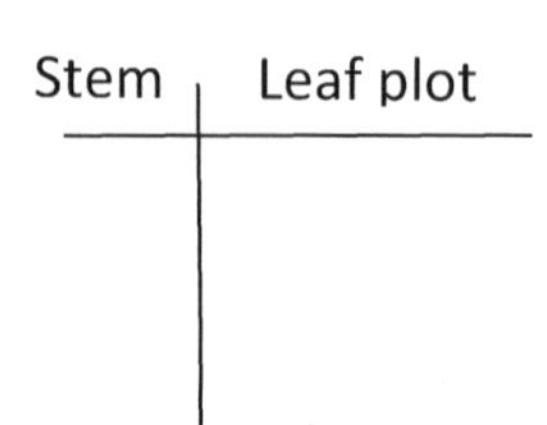

3) 58, 41, 42, 67, 54, 65, 65, 54, 69, 53

Stem | Leaf plot

The Pie Graph or Circle Graph

Helpful Hints

A Pie Chart is a circle chart divided into sectors, each sector represents the relative size of each value.

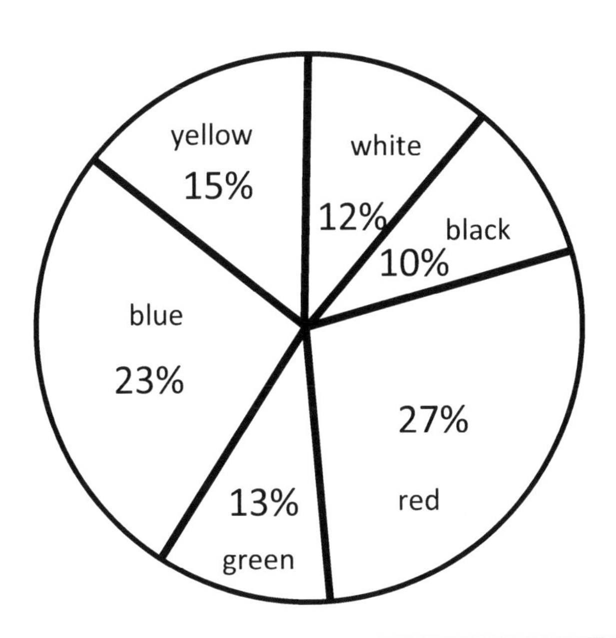

Favorite colors

1) Which color is the most?
2) What percentage of pie graph is yellow?
3) Which color is the least?
4) What percentage of pie graph is blue?
5) What percentage of pie graph is green?

Scatter Plots

Helpful Hints

A Scatter (xy) Plot shows the values with points that represent the relationship between two sets of data.

– The horizontal values are usually x and vertical data is y.

Construct a scatter plot.

X	Y
1	20
2	40
3	50
4	60

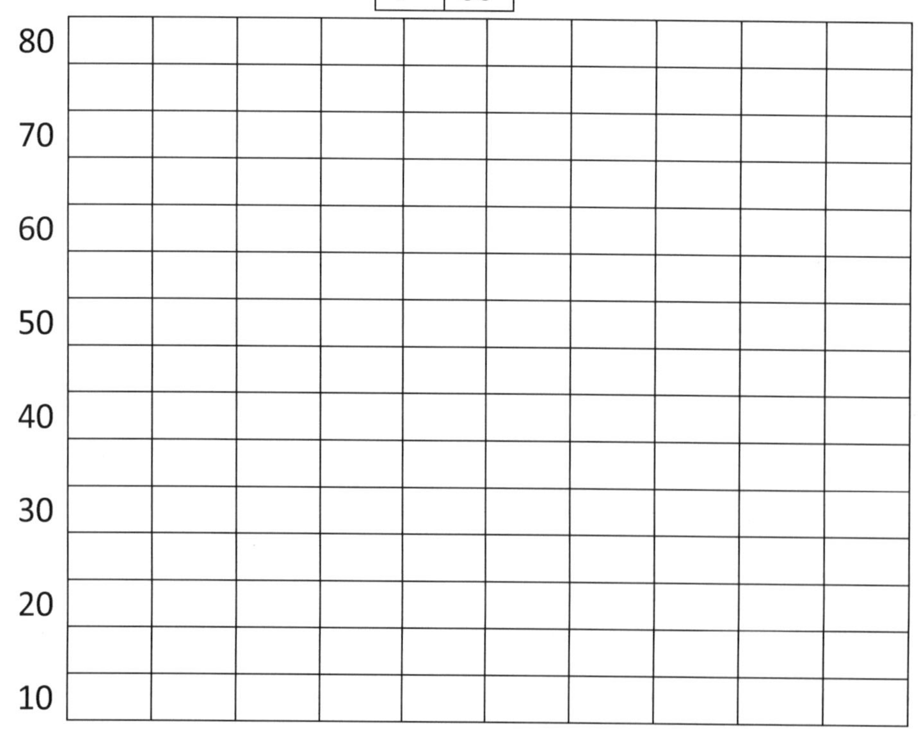

Probability Problems

Helpful Hints		Example:
	- Probability is the likelihood of something happening in the future. It is expressed as a number between zero (can never happen) to 1 (will always happen). - Probability can be expressed as a fraction, a decimal, or a percent.	Probability of a flipped coins turns up 'heads' Is $0.5 = \frac{1}{2}$

Solve.

1) A number is chosen at random from 1 to 10. Find the probability of selecting a 4 or smaller.

2) A number is chosen at random from 1 to 50. Find the probability of selecting multiples of 10.

3) A number is chosen at random from 1 to 10. Find the probability of selecting of 4 and factors of 6.

4) A number is chosen at random from 1 to 10. Find the probability of selecting a multiple of 3.

5) A number is chosen at random from 1 to 50. Find the probability of selecting prime numbers.

6) A number is chosen at random from 1 to 25. Find the probability of not selecting a composite number.

Answers of Worksheets – Chapter 15

Mean, Median, Mode, and Range of the Given Data

1) mean: 3, median: 2, mode: 1, 2, range: 6
2) mean: 3.625, median: 3, mode: 2, range: 5
3) mean: 5.22, median: 4, mode: 4, range: 8
4) mean: 4, median: 4, mode: 4, range: 6
5) mean: 7, median: 7, mode: 5, 7, 8, range: 4
6) mean: 4.2, median: 4, mode: 1,2,4,5,9, range: 8
7) mean: 5, median: 5, mode: 5, range: 8
8) mean: 5.78, median: 6, mode: 7, range: 7
9) mean: 5, median: 5, mode: 2, 4, 5, 6, range: 7
10) mean: 6.125, median: 5, mode: 5, range: 8
11) mean: 3, median: 2, mode: 2, 5, range: 4
12) mean: 5, median: 5, mode: none, range: 7

Box and Whisker Plots

11, 17, 22, 18, 23, 2, 3, 16, 21, 7, 8, 15, 5

Maximum: 23, Minimum: 2, Q_1: 2, Q_2: 12.5, Q_3: 19.5

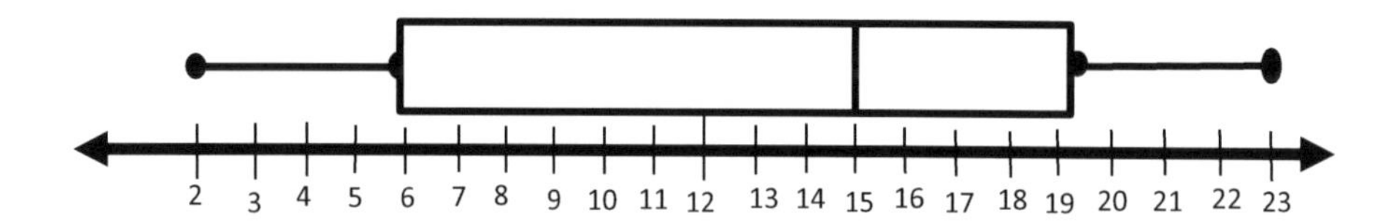

Bar Graph

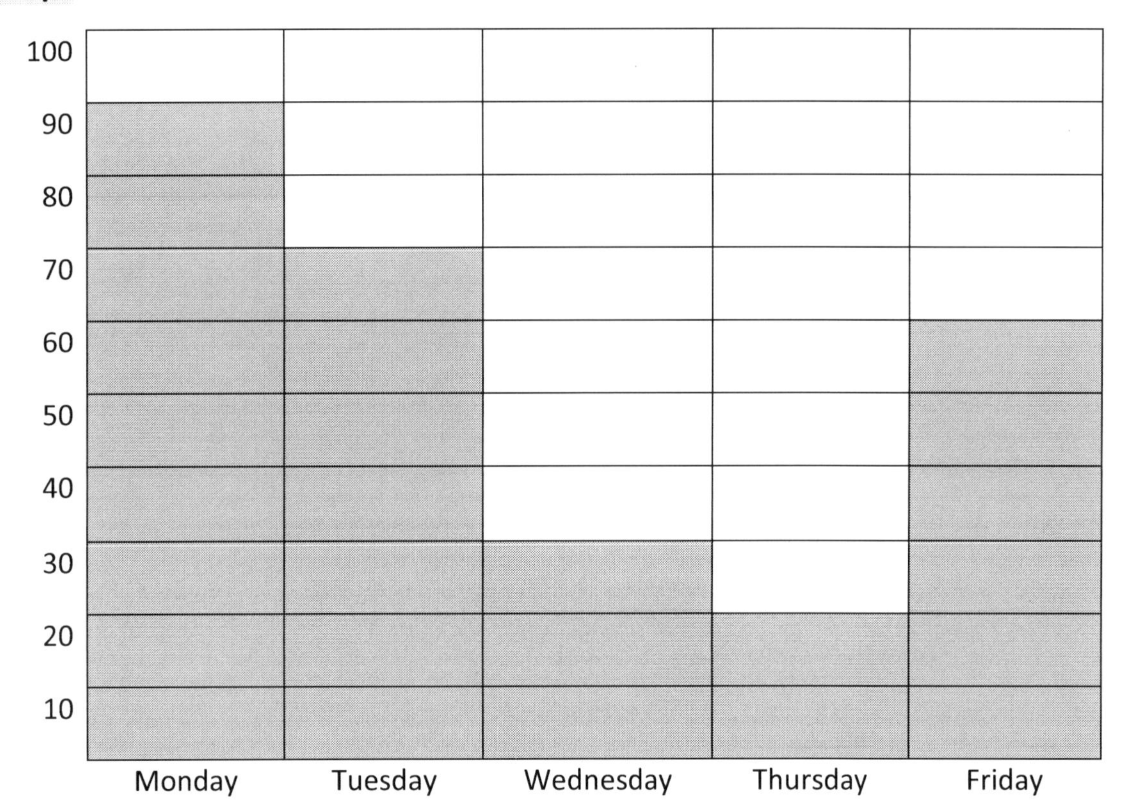

Stem–And–Leaf Plot

1)

Stem	leaf
7	2 4 9 9
8	3 6 8
9	1 5 7

2)

Stem	leaf
1	9
2	6 6 6
3	3 7
4	8 8 9

3)

Stem	leaf
4	1 2
5	3 4 4 8
6	5 5 7 9

The Pie Graph or Circle Graph

1) red
2) 15%
3) black
4) 23%
5) 13%

Scatter Plots

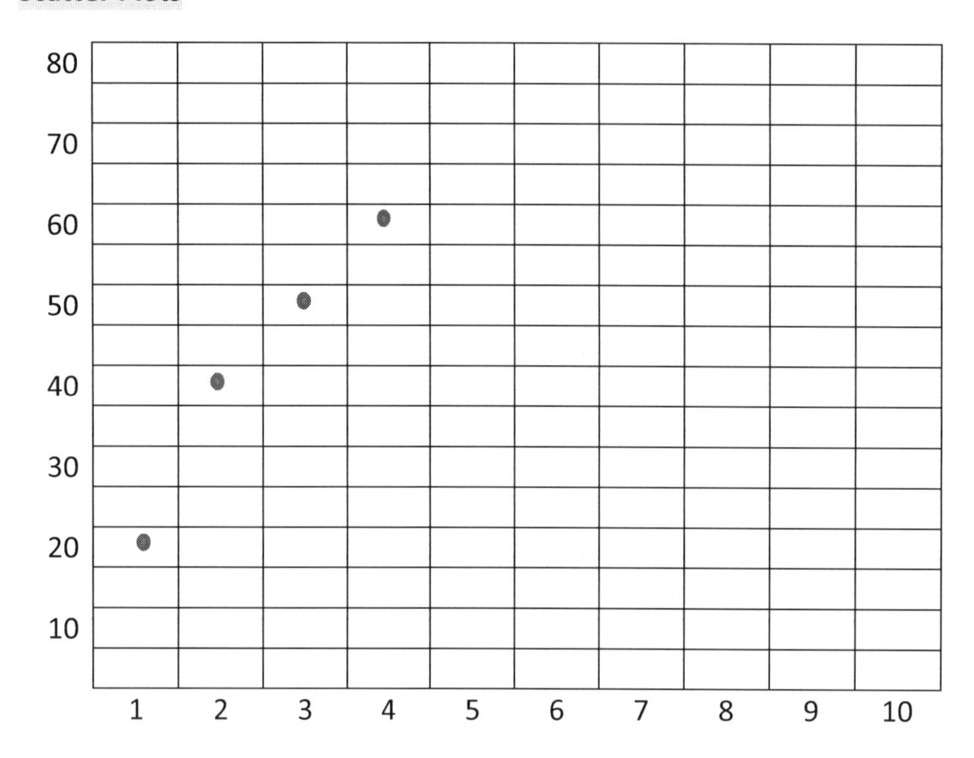

Probability Problems

1) $\frac{2}{5}$
2) $\frac{1}{10}$
3) $\frac{1}{2}$
4) $\frac{3}{10}$
5) $\frac{7}{25}$
6) $\frac{9}{25}$

SSAT Upper Level Test Review

The SSAT, or Secondary School Admissions Test, is a standardized test to help determine admission to private elementary, middle and high schools.

There are currently three Levels of the SSAT:

- ✓ Lower Level (for students in 3rd and 4th grade)
- ✓ Middle Level (for students in 5th-7th grade)
- ✓ Upper Level (for students in 8th-11th grade)

The SSAT Upper Level test consists of three separate exams:

- ✓ Quantitative (Mathematics)
- ✓ Reading
- ✓ Writing

There are two quantitative sections on the test. The mathematics portions of the SSAT Upper Level test each contains 25 questions. For each math section, students have 30 minutes to complete the test. The Math section of the test covers arithmetic, data analysis, geometry, algebra, and some basic statistics topics.
Students are not allowed to use calculator when taking a SSAT Upper Level assessment.
In this section, there are two complete SSAT Upper Level Mathematics Tests. Take these tests to see what score you'll be able to receive on a real SSAT Upper Level test.

Good luck!

SSAT Upper Level Math Practice Tests

Time to Test

Time to refine your skill with a practice examination

Take practice SSAT Upper Level Math Tests to simulate the test day experience. After you've finished, score your tests using the answer keys.

Before You Start

- You'll need a pencil and a timer to take the test.
- After you've finished the test, review the answer key to see where you went wrong.
- Use the answer sheet provided to record your answers. (You can cut it out or photocopy it)
- You will receive 1 point for every correct answer and you will lose $\frac{1}{4}$ point for each incorrect answer. There is no penalty for skipping a question.

Calculators are NOT permitted for the SSAT Middle Level Test

Good Luck!

SSAT Practice Test 1 Answer Sheet

Remove (or photocopy) this answer sheet and use it to complete the practice test.

SSAT Upper Level Mathematics Practice Test 1 Answer Sheet

SSAT Upper Level Practice Test 1 Section 1

1	(A) (B) (C) (D) (E)	11	(A) (B) (C) (D) (E)	21	(A) (B) (C) (D) (E)
2	(A) (B) (C) (D) (E)	12	(A) (B) (C) (D) (E)	22	(A) (B) (C) (D) (E)
3	(A) (B) (C) (D) (E)	13	(A) (B) (C) (D) (E)	23	(A) (B) (C) (D) (E)
4	(A) (B) (C) (D) (E)	14	(A) (B) (C) (D) (E)	24	(A) (B) (C) (D) (E)
5	(A) (B) (C) (D) (E)	15	(A) (B) (C) (D) (E)	25	(A) (B) (C) (D) (E)
6	(A) (B) (C) (D) (E)	16	(A) (B) (C) (D) (E)		
7	(A) (B) (C) (D) (E)	17	(A) (B) (C) (D) (E)		
8	(A) (B) (C) (D) (E)	18	(A) (B) (C) (D) (E)		
9	(A) (B) (C) (D) (E)	19	(A) (B) (C) (D) (E)		
10	(A) (B) (C) (D) (E)	20	(A) (B) (C) (D) (E)		

SSAT Upper Level Practice Test 1 Section 2

1	(A) (B) (C) (D) (E)	11	(A) (B) (C) (D) (E)	21	(A) (B) (C) (D) (E)
2	(A) (B) (C) (D) (E)	12	(A) (B) (C) (D) (E)	22	(A) (B) (C) (D) (E)
3	(A) (B) (C) (D) (E)	13	(A) (B) (C) (D) (E)	23	(A) (B) (C) (D) (E)
4	(A) (B) (C) (D) (E)	14	(A) (B) (C) (D) (E)	24	(A) (B) (C) (D) (E)
5	(A) (B) (C) (D) (E)	15	(A) (B) (C) (D) (E)	25	(A) (B) (C) (D) (E)
6	(A) (B) (C) (D) (E)	16	(A) (B) (C) (D) (E)		
7	(A) (B) (C) (D) (E)	17	(A) (B) (C) (D) (E)		
8	(A) (B) (C) (D) (E)	18	(A) (B) (C) (D) (E)		
9	(A) (B) (C) (D) (E)	19	(A) (B) (C) (D) (E)		
10	(A) (B) (C) (D) (E)	20	(A) (B) (C) (D) (E)		

SSAT Upper Level Math

Practice Test 1

Section 1

25 questions

Total time for this section: 30 Minutes

You may NOT use a calculator for this test.

1) A shaft rotates 300 times in 8 seconds. How many times does it rotate in 12 seconds?

A. 450
B. 300
C. 200
D. 150
E. 100

2) A school wants to give each of its 20 top students a football ball. If the balls are in boxes of three, how many boxes of balls they need to purchase?

A. 3
B. 5
C. 6
D. 7
E. 20

3) If $\frac{25}{A} + 1 = 6$, then $25 + A = ?$

A. 6
B. 1
C. 25
D. 30
E. 0

4) How many tiles of 8 cm^2 is needed to cover a floor of dimension 6 cm by 24 cm?

A. 6

B. 12

C. 18

D. 24

E. 36

5) Which of the following statements is correct, according to the graph below?

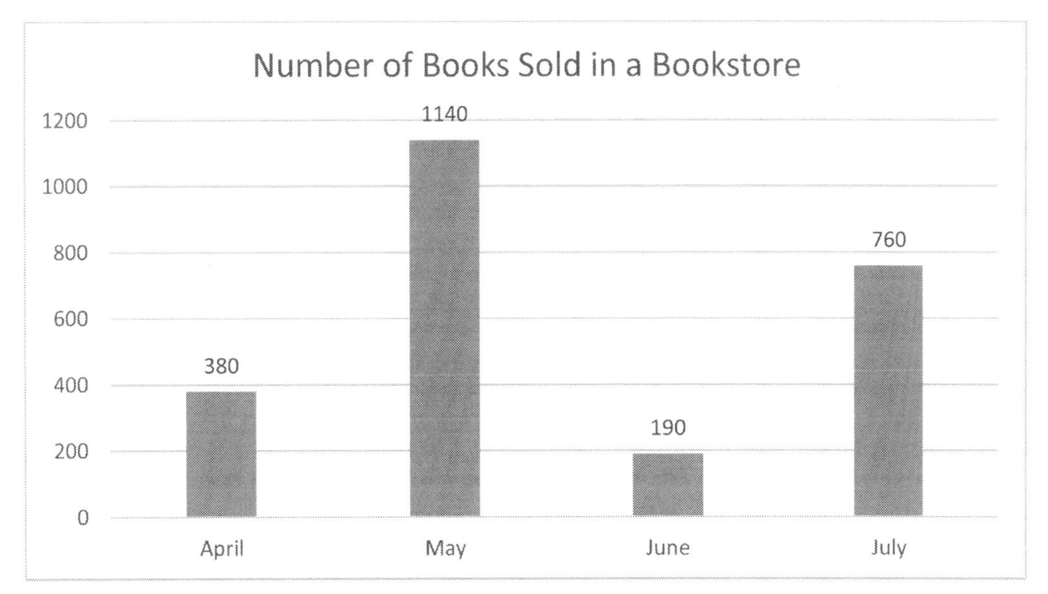

A. Number of books sold in April was twice the number of books sold in July.

B. Number of books sold in July was less than half the number of books sold in May.

C. Number of books sold in June was half the number of books sold in April.

D. Number of books sold in July was equal to the number of books sold in April plus the number of books sold in June.

E. More books were sold in April than in July.

6) $12.124 \div 0.002$?

A. 6.0620

B. 60.620

C. 606.20

D. 6,062.0

E. 600620

7) What is the value of the sum of the tens and thousandths in number 2,517.89451?

A. 16

B. 11

C. 9

D. 14

E. 5

8) If 150 % of a number is 75, then what is the 90 % of that number?

A. 45

B. 50

C. 60

D. 70

E. 85

9) Jack earns \$616 for his first 44 hours of work in a week and is then paid 1.5 times his regular hourly rate for any additional hours. This week, Jack needs \$826 to pay his rent, bills and other expenses. How many hours must he work to make enough money in this week?

A. 40
B. 48
C. 50
D. 53
E. 54

10) $\frac{1\frac{3}{4}+\frac{1}{3}}{2\frac{1}{2}-\frac{15}{8}}$ is approximately equal to.

A. 3.33
B. 3.6
C. 5.67
D. 6.33
E. 6.67

11) If $1 \leq x < 4$, what is the minimum value of the following expression?

$$2x + 1$$

A. 8
B. 5
C. 3
D. 2
E. 1

12) If $x■y = \sqrt{x^2 + y}$, what is the value of $5■11$?

A. $\sqrt{126}$
B. 6
C. 4
D. 3
E. 2

13) The average weight of 18 girls in a class is 60 kg and the average weight of 32 boys in the same class is 62 kg. What is the average weight of all the 50 students in that class?

A. 60
B. 61.28
C. 61.68
D. 61.9
E. 62.20

14) There are three equal tanks of water. If 2/5 of a tank contains 200 liters of water, what is the capacity of the three tanks of water together?

A. 1,500
B. 500
C. 240
D. 80
E. 200

15) What is the answer of $7.5 \div 0.15$?

A. $\frac{1}{50}$

B. $\frac{1}{5}$

C. 5

D. 50

E. 500

16) Two-kilograms apple and three-kilograms orange cost \$26.4. If one-kilogram apple costs \$4.2 how much does one-kilogram orange cost?

A. \$9

B. \$6

C. \$5.5

D. \$5

E. \$4

17) What is the value of x in the following figure?

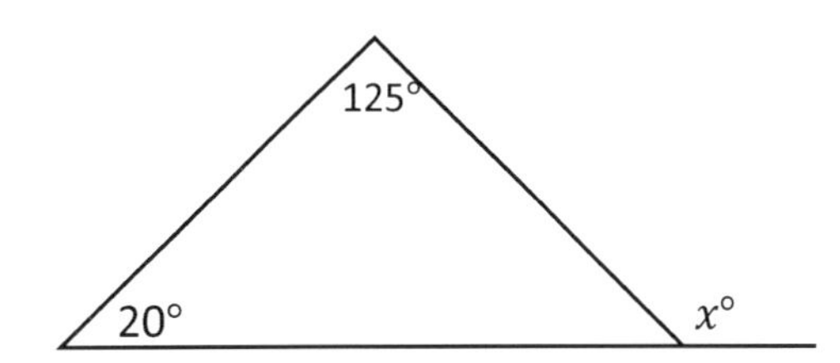

A. 150

B. 145

C. 125

D. 115

E. 105

18) David's current age is 42 years, and Ava's current age is 6 years old. In how many years David's age will be 4 times Ava's age?

A. 4
B. 6
C. 8
D. 10
E. 14

19) Michelle and Alec can finish a job together in 100 minutes. If Michelle can do the job by herself in 5 hours, how many minutes does it take Alec to finish the job?

A. 120
B. 150
C. 180
D. 200
E. 220

20) The sum of six different negative integers is -70. If the smallest of these integers is -15, what is the largest possible value of one of the other five integers?

A. -14
B. -10
C. -5
D. -4
E. -1

21) What is the slope of a line that is perpendicular to the line $4x - 2y = 12$?

A. -2

B. $-\frac{1}{2}$

C. 4

D. 12

E. 14

22) A cruise line ship left Port A and traveled 80 miles due west and then 150 miles due north. At this point, what is the shortest distance from the cruise to port A?

A. 70 miles

B. 80 miles

C. 150 miles

D. 170 miles

E. 230 miles

23) The Jackson Library is ordering some bookshelves. If x is the number of bookshelf the library wants to order, which each costs $100 and there is a one-time delivery charge of $800, which of the following represents the total cost, in dollar, per bookshelf?

A. $100x + 800$

B. $100 + 800x$

C. $\frac{100x+800}{100}$

D. $\frac{100x+800}{x}$

E. $100x - 800$

24) A football team won exactly 80% of the games it played during last session. Which of the following could be the total number of games the team played last season?

A. 49

B. 35

C. 32

D. 16

E. 12

25) The width of a box is one third of its length. The height of the box is one third of its width. If the length of the box is 27 cm, what is the volume of the box?

A. 81 cm^3

B. 162 cm^3

C. 243 cm^3

D. 729 cm^3

E. 1880 cm^3

IF YOU FINISH BEFORE TIME IS CALLED, YOU MAY CHECK YOUR WORK ON THIS SECTION ONLY. DO NOT TURN TO ANY OTHER SECTION IN THE TEST.

STOP

SSAT Upper Level Math

Test 1 Section 2

25 questions

Total time for this section: 30 Minutes

You may NOT use a calculator for this test.

1) There are 11 marbles in the bag A and 17 marbles in the bag B. If the sum of the marbles in both bags will be shared equally between two children, how many marbles bag A has less than the marbles that each child will receive?

A. 2
B. 3
C. 4
D. 5
E. 6

2) If the perimeter of the following figure be 20, what is the value of x?

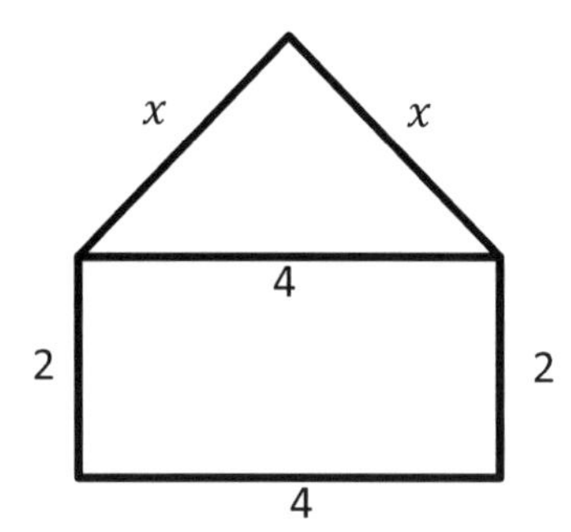

A. 2
B. 3
C. 6
D. 9
E. 12

3) When number 91,501 is divided by 305, the result is closest to?

A. 3
B. 30
C. 300
D. 350
E. 400

4) If Jason's mark is k more than Alex, and Jason's mark is 16, which of the following can be Alex's mark?

A. $16 + k$
B. $k - 16$
C. $\frac{k}{16}$
D. $16k$
E. $16 - k$

5) To paint a wall with the area of $36m^2$, how many liters of paint do we need if each liter of paint is enough to paint a wall with dimension of $72\ cm \times 100\ cm$?

A. 50
B. 100
C. 150
D. 200
E. 250

6) The price of a sofa is decreased by 25% to $420. What was its original price?

A. $480
B. $520
C. $560
D. $600
E. $800

7) $750 - 7\frac{7}{15} = ?$

A. $742\frac{7}{15}$

B. $742\frac{8}{15}$

C. $743\frac{1}{15}$

D. $743\frac{8}{15}$

E. $744\frac{1}{15}$

8) A driver rests one hour and 12 minutes for every 4 hours driving. How many minutes will he rest if he drives 20 hours?

A. 3 hours and 36 minutest

B. 4 hours and 12 minutest

C. 4 hours and 45 minutest

D. 5 hours and 36 minutest

E. 6 hours

9) Which of the following expression is not equal to 5?

A. $10 \times \frac{1}{2}$

B. $25 \times \frac{1}{5}$

C. $2 \times \frac{5}{2}$

D. $6 \times \frac{5}{6}$

E. $5 \times \frac{1}{5}$

10) What is the missing term in the given sequence?

2, 3, 5, 8, 12, 17, 23, ___, 38

A. 24

B. 26

C. 27

D. 28

E. 30

Questions 11 to 12 are based on the following graph

A library has 840 books that include Mathematics, Physics, Chemistry, English and History.

Use following graph to answer questions 11 to 12.

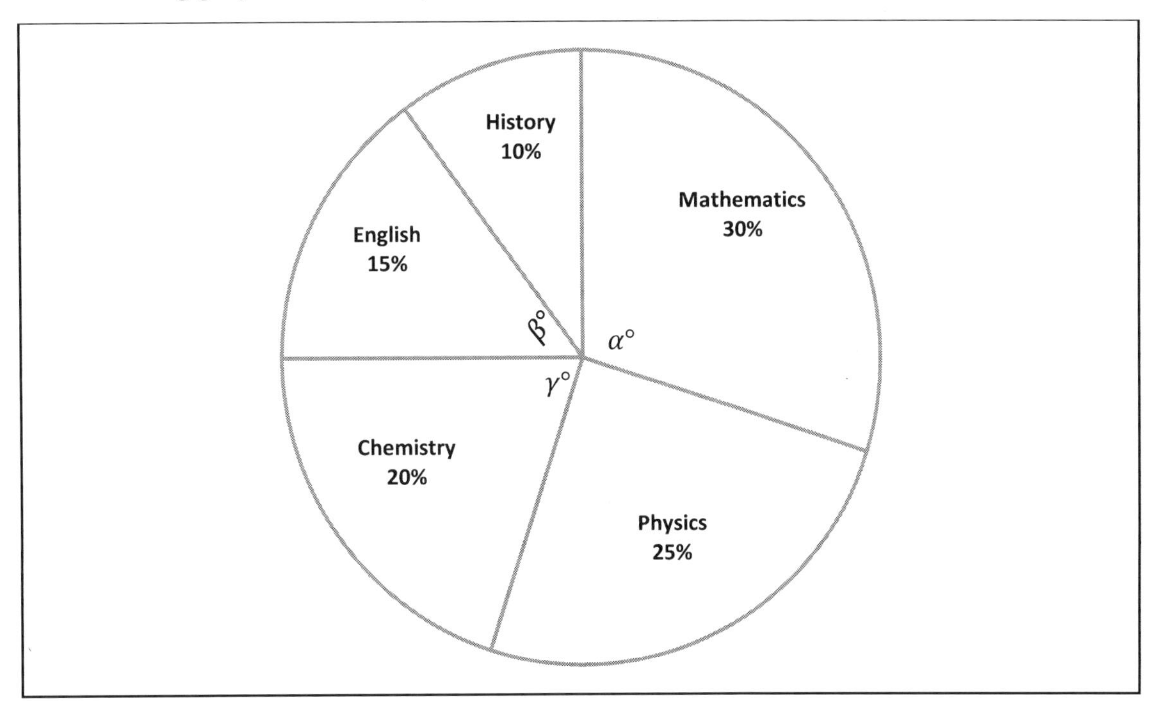

11) What is the product of the number of Mathematics and number of English books?

A. 21168

B. 31752

C. 26460

D. 17640

E. 35280

12) What are the values of angle α and β respectively?

A. $90^\circ, 54^\circ$

B. $120^\circ, 36^\circ$

C. $120^\circ, 45^\circ$

D. $108^\circ, 54^\circ$

E. $108^\circ, 45^\circ$

13) Find the perimeter of following shape.

A. 21

B. 22

C. 23

D. 24

E. 25

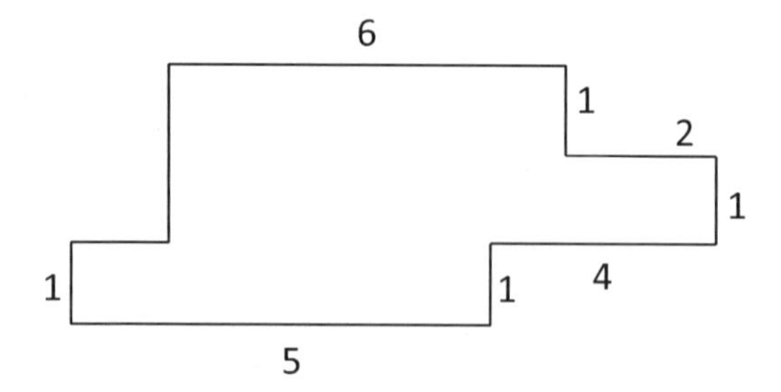

14) If $3y + 5 < 29$, then y could be equal to?

A. 15

B. 12

C. 10.5

D. 8

E. 2.5

15) The capacity of a red box is 20% bigger than the capacity of a blue box. If the red box can hold 30 equal sized books, how many of the same books can the blue box hold?

A. 9

B. 15

C. 21

D. 25

E. 30

16) If the area of the following rectangular ABCD is 100, and E is the midpoint of AB, what is the area of the shaded part?

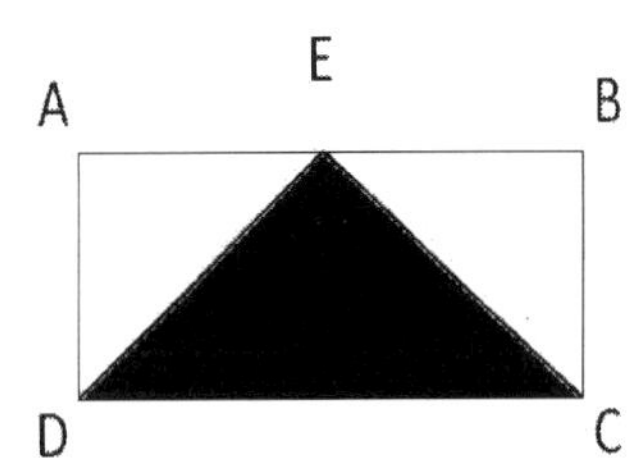

A. 25

B. 50

C. 75

D. 80

E. 100

17) There are 50.2 liters of gas in a car fuel tank. In the first week and second week of April, the car uses 5.82 and 25.9 liters of gas respectively. If the car was park in the third week of April and 10.31 liters of gas will be added to the fuel tank, how many liters of gas are in the fuel tank of the car?

A. 21.41
B. 25.9
C. 27
D. 28.79
E. 71.61

18) If $3x + y = 25$ and $x - z = 14$, what is the value of x?

A. 0
B. 5
C. 10
D. 20
E. it cannot be determined from the information given

19) If $a \times b$ is divisible by 3, which of the following expression must also be divisible by 3?

A. 0
B. $3a - b$
C. $a - 3b$
D. $\frac{a}{b}$
E. $4 \times a \times b$

20) What is the average of circumference of figure A and area of figure B? ($\pi = 3$)

A. 54

B. 53

C. 52

D. 51

E. 50

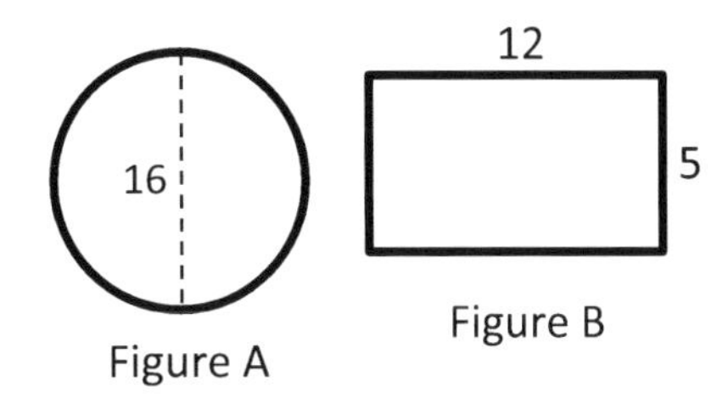

21) In the following figure, point Q lies on line n, what is the value of y if $x = 35$?

A. 15

B. 25

C. 35

D. 45

E. 55

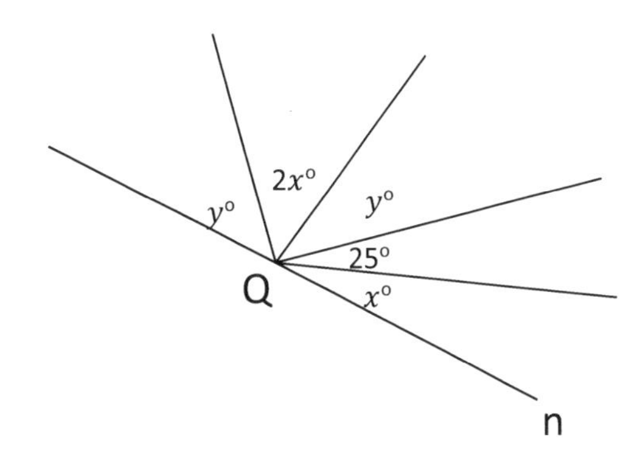

22) Which of the following could be the value of x if $\frac{5}{9} + x > 2$?

A. $\frac{1}{2}$

B. $\frac{3}{5}$

C. $\frac{4}{5}$

D. $\frac{4}{3}$

E. $\frac{5}{3}$

23) A number is chosen at random from 1 to 25. Find the probability of not selecting a composite number. (A composite number is a number that is divisible by itself, 1 and at least one other whole number)

A. $\frac{1}{25}$

B. $\frac{2}{5}$

C. $\frac{9}{25}$

D. 1

E. 0

24) $712 \div 3 =?$

A. $\frac{700}{3} \times \frac{10}{3} \times \frac{2}{3}$

B. $700 + \frac{10}{3} + \frac{2}{3}$

C. $\frac{700}{3} + \frac{10}{3} + \frac{2}{3}$

D. $\frac{700}{3} \div \frac{10}{3} \div \frac{2}{3}$

E. $\frac{7}{3} + \frac{1}{3} + \frac{2}{3}$

25) If a gas tank can hold 25 gallons, how many gallons does it contain when it is $\frac{2}{5}$ full?

A. 125

B. 62.5

C. 50

D. 10

E. 5

IF YOU FINISH BEFORE TIME IS CALLED, YOU MAY CHECK YOUR WORK ON THIS SECTION ONLY. DO NOT TURN TO ANY OTHER SECTION IN THE TEST.

STOP

SSAT Practice Test 2 Answer Sheet

Remove (or photocopy) this answer sheet and use it to complete the practice test.

SSAT Upper Level Mathematics Practice Test 2 Answer Sheet

SSAT Upper Level Practice Test 2 Section 1

1	Ⓐ Ⓑ Ⓒ Ⓓ Ⓔ	11	Ⓐ Ⓑ Ⓒ Ⓓ Ⓔ	21	Ⓐ Ⓑ Ⓒ Ⓓ Ⓔ
2	Ⓐ Ⓑ Ⓒ Ⓓ Ⓔ	12	Ⓐ Ⓑ Ⓒ Ⓓ Ⓔ	22	Ⓐ Ⓑ Ⓒ Ⓓ Ⓔ
3	Ⓐ Ⓑ Ⓒ Ⓓ Ⓔ	13	Ⓐ Ⓑ Ⓒ Ⓓ Ⓔ	23	Ⓐ Ⓑ Ⓒ Ⓓ Ⓔ
4	Ⓐ Ⓑ Ⓒ Ⓓ Ⓔ	14	Ⓐ Ⓑ Ⓒ Ⓓ Ⓔ	24	Ⓐ Ⓑ Ⓒ Ⓓ Ⓔ
5	Ⓐ Ⓑ Ⓒ Ⓓ Ⓔ	15	Ⓐ Ⓑ Ⓒ Ⓓ Ⓔ	25	Ⓐ Ⓑ Ⓒ Ⓓ Ⓔ
6	Ⓐ Ⓑ Ⓒ Ⓓ Ⓔ	16	Ⓐ Ⓑ Ⓒ Ⓓ Ⓔ		
7	Ⓐ Ⓑ Ⓒ Ⓓ Ⓔ	17	Ⓐ Ⓑ Ⓒ Ⓓ Ⓔ		
8	Ⓐ Ⓑ Ⓒ Ⓓ Ⓔ	18	Ⓐ Ⓑ Ⓒ Ⓓ Ⓔ		
9	Ⓐ Ⓑ Ⓒ Ⓓ Ⓔ	19	Ⓐ Ⓑ Ⓒ Ⓓ Ⓔ		
10	Ⓐ Ⓑ Ⓒ Ⓓ Ⓔ	20	Ⓐ Ⓑ Ⓒ Ⓓ Ⓔ		

SSAT Upper Level Practice Test 2 Section 2

1	Ⓐ Ⓑ Ⓒ Ⓓ Ⓔ	11	Ⓐ Ⓑ Ⓒ Ⓓ Ⓔ	21	Ⓐ Ⓑ Ⓒ Ⓓ Ⓔ
2	Ⓐ Ⓑ Ⓒ Ⓓ Ⓔ	12	Ⓐ Ⓑ Ⓒ Ⓓ Ⓔ	22	Ⓐ Ⓑ Ⓒ Ⓓ Ⓔ
3	Ⓐ Ⓑ Ⓒ Ⓓ Ⓔ	13	Ⓐ Ⓑ Ⓒ Ⓓ Ⓔ	23	Ⓐ Ⓑ Ⓒ Ⓓ Ⓔ
4	Ⓐ Ⓑ Ⓒ Ⓓ Ⓔ	14	Ⓐ Ⓑ Ⓒ Ⓓ Ⓔ	24	Ⓐ Ⓑ Ⓒ Ⓓ Ⓔ
5	Ⓐ Ⓑ Ⓒ Ⓓ Ⓔ	15	Ⓐ Ⓑ Ⓒ Ⓓ Ⓔ	25	Ⓐ Ⓑ Ⓒ Ⓓ Ⓔ
6	Ⓐ Ⓑ Ⓒ Ⓓ Ⓔ	16	Ⓐ Ⓑ Ⓒ Ⓓ Ⓔ		
7	Ⓐ Ⓑ Ⓒ Ⓓ Ⓔ	17	Ⓐ Ⓑ Ⓒ Ⓓ Ⓔ		
8	Ⓐ Ⓑ Ⓒ Ⓓ Ⓔ	18	Ⓐ Ⓑ Ⓒ Ⓓ Ⓔ		
9	Ⓐ Ⓑ Ⓒ Ⓓ Ⓔ	19	Ⓐ Ⓑ Ⓒ Ⓓ Ⓔ		
10	Ⓐ Ⓑ Ⓒ Ⓓ Ⓔ	20	Ⓐ Ⓑ Ⓒ Ⓓ Ⓔ		

SSAT Upper Level Math

Practice Test 2

Section 1

25 questions

Total time for this section: 30 Minutes

You may NOT use a calculator for this test.

1) $0.03 \times 12.00 = ?$

A. 3.6

B. 36.00

C. 0.36

D. 3.06

E. 0.036

2) If $x - 10 = -10$, then $x \times 3 = ?$

A. 10

B. 30

C. 60

D. 90

E. 0

3) What is the value of the "4" in number 131.493?

A. 4 ones

B. 4 tenths

C. 4 hundredths

D. 4 tens

E. 4 thousandths

4) Mia plans to buy a bracelet for every one of her 16 friends for their party. There are three bracelets in each pack. How many packs must she buy?

A. 3
B. 4
C. 5
D. 6
E. 10

5) If Logan ran 2.5 miles in half an hour, his average speed was?

A. 1.25 miles per hour
B. 2.5 miles per hour
C. 3.75 miles per hour
D. 5 miles per hour
E. 10 miles per hour

6) A pizza maker has x pounds of flour to make pizzas. After he has used 55 pounds of flour, how much flour is left? The expression that correctly represents the quantity of flour left is:

A. $55 + x$
B. $\frac{55}{x}$
C. $55 - x$
D. $x - 55$
E. $55x$

7) The distance between cities A and B is approximately 2,600 miles. If Alice drives an average of 68 miles per hour, how many hours will it take Alice to drive from city A to city B?

A. Approximately 41 hours
B. Approximately 38 hours
C. Approximately 29 hours
D. Approximately 27 hours
E. Approximately 21 hours

8) Given the diagram, what is the perimeter of the quadrilateral?

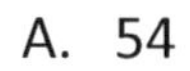

A. 54
B. 66
C. 620
D. 16740
E. 33480

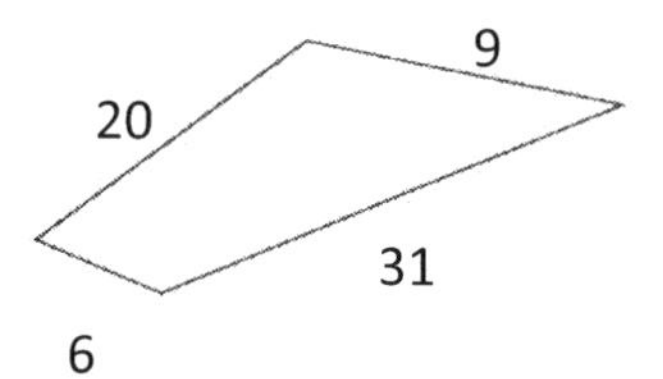

9) In a classroom of 60 students, 42 are female. What percentage of the class is male?

A. 34%
B. 22%
C. 30%
D. 26%
E. 15%

10) An employee's rating on performance appraisals for the last three quarters were 92, 88 and 86. If the required yearly average to qualify for the promotion is 90, what rating should the fourth quarter be?

A. 91
B. 92
C. 93
D. 94
E. 95

11) Two third of 18 is equal to $\frac{2}{5}$ of what number?

A. 12
B. 20
C. 30
D. 60
E. 90

12) A steak dinner at a restaurant costs $8.25. If a man buys a steak dinner for himself and 3 friends, what will the total cost be?

A. $33
B. $17.01
C. $27
D. $21.5
E. 11

13) Last week 24,000 fans attended a football match. This week three times as many bought tickets, but one sixth of them cancelled their tickets. How many are attending this week?

A. 48,000

B. 54,000

C. 60,000

D. 72,000

E. 84,000

14) What is the slope of the line that is perpendicular to the line with equation $7x + y = 12$?

A. $\frac{1}{7}$

B. $-\frac{1}{7}$

C. $\frac{7}{12}$

D. 7

E. -7

15) A cruise line ship left Port A and traveled 50 miles due west and then 120 miles due north. At this point, what is the shortest distance from the cruise to port A?

A. 70 miles

B. 80 miles

C. 130 miles

D. 160 miles

E. 170 miles

16) In 1999, the average worker's income increased \$2,000 per year starting from \$24,000 annual salary. Which equation represents income greater than average? (I = income, x = number of years after 1999)

A. $I > 2000x + 24000$
B. $I > -2000x + 24000$
C. $I < -2000x + 24000$
D. $I < 2000x - 24000$
E. $I < 24,000x + 24000$

Questions 17 to 19 are based on the following data

The result of a research shows the number of men and women in four cities of a country.

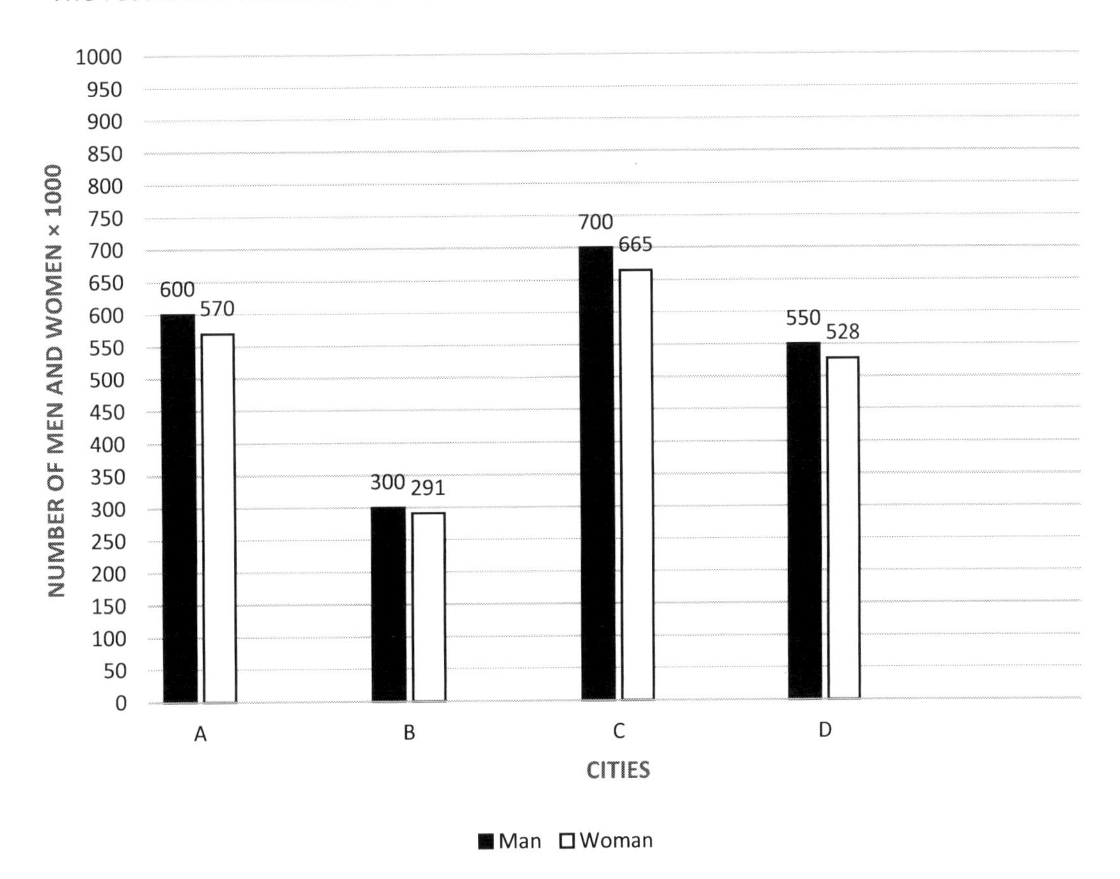

17) What's the maximum ratio of woman to man in the four cities?

A. 0.98
B. 0.97
C. 0.96
D. 0.95
E. 0.93

18) What's the ratio of percentage of men in city A to percentage of women in city C?

A. 0.9
B. 0.95
C. 1
D. 1.05
E. 1.5

19) How many women should be added to city D until the ratio of women to men will be 1.2?

A. 120
B. 128
C. 132
D. 160
E. 165

20) In the following figure, MN is 40 cm. How long is ON?

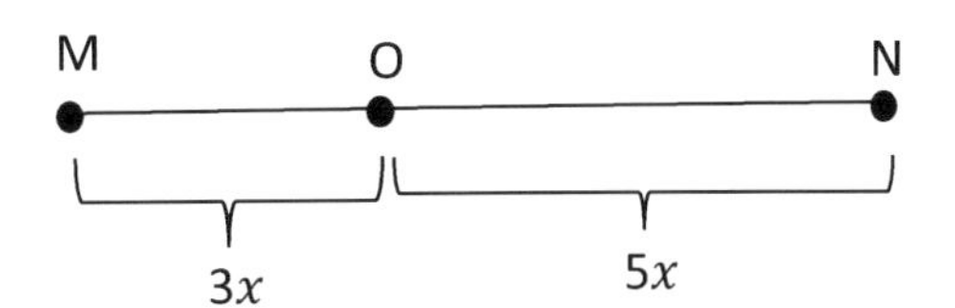

A. 25 cm

B. 20 cm

C. 15 cm

D. 10 cm

E. 5 cm

21) Solve the following equation for y?

$$\frac{x}{2+3}=\frac{y}{10-7}$$

A. $\frac{3}{5}x$

B. $\frac{5}{3}x$

C. $3x$

D. $2x$

E. $\frac{1}{2}x$

22) $\sqrt[5]{x^{16}}=?$

A. $x^3\sqrt[5]{x}$

B. $80x$

C. x^{11}

D. x^{80}

E. x^4

23) John traveled 150 km in 6 hours and Alice traveled 180 km in 4 hours. What is the ratio of the average speed of John to average speed of Alice?

A. 3 : 2

B. 2 : 3

C. 5 : 9

D. 5 : 6

E. 11 : 16

24) What is the difference of smallest 4–digit number and biggest 4–digit number?

A. 6,666

B. 6,789

C. 8,888

D. 8,999

E. 9,999

25) Rectangle A has a length of 8 cm and a width of 4cm, and rectangle B has a length of 5 cm and a width of 4 cm, what is the percent of ratio of the perimeter of rectangle B to rectangle A?

A. 15%

B. 25%

C. 50%

D. 75%

E. 133.3%

IF YOU FINISH BEFORE TIME IS CALLED, YOU MAY CHECK YOUR WORK ON THIS SECTION ONLY. DO NOT TURN TO ANY OTHER SECTION IN THE TEST.

STOP

SSAT Upper Level Math

Test 2 Section 2

25 questions

Total time for this section: 30 Minutes

You may NOT use a calculator for this test.

1) $5\frac{3}{7} \times 4\frac{1}{5} = ?$

A. $23\frac{1}{5}$

B. $23\frac{4}{5}$

C. $22\frac{4}{5}$

D. $22\frac{1}{5}$

E. 21

2) Which of the following is a whole number ?

A. $\frac{2}{3} \times \frac{9}{5}$

B. $\frac{1}{2} + \frac{1}{4}$

C. $\frac{21}{6}$

D. $2.5 + 1$

E. $2.5 + \frac{7}{2}$

3) $0.42 \times 11.8 = ?$

A. 4.956

B. 4.965

C. 5.956

D. 5.965

E. 5.695

4) Sophia purchased a sofa for $530.40. The sofa is regularly priced at $624. What was the percent discount Sophia received on the sofa?

A. 12%

B. 15%

C. 20%

D. 25%

E. 40%

5) If $\frac{2}{5}$ of a number equal to 12 then $\frac{2}{3}$ of the same number is:

A. 48

B. 45

C. 30

D. 24

E. 20

6) How long does a 420–miles trip take moving at 50 miles per hour (mph)?

A. 6 hours

B. 6 hours and 24 minutes

C. 8 hours and 24 minutes

D. 8 hours and 30 minutes

E. 10 hours and 30 minutes

7) A swimming pool holds 2,000 cubic feet of water. The swimming pool is 25 feet long and 10 feet wide. How deep is the swimming pool?

A. 2 feet
B. 4 feet
C. 6 feet
D. 7 feet
E. 8 feet

8) A bank is offering 3.5% simple interest on a savings account. If you deposit $12,000, how much interest will you earn in two years?

A. $420
B. $840
C. $4,200
D. $8,400
E. $9,000

9) In the figure below, line A is parallel to line B. What is the value of angle x?

A. 35 degree

B. 45 degree

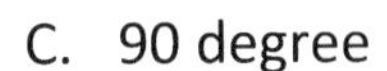
C. 90 degree

D. 100 degree

E. 145 degree

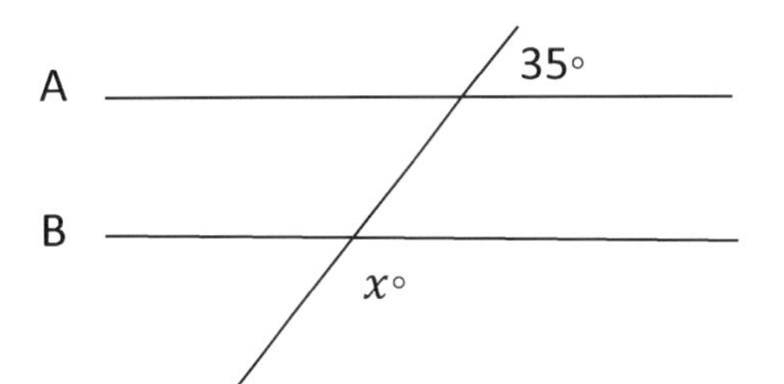

10) We can put 24 colored pencils in each box and we have 408 colored pencils. How many boxes do we need?

A. 13
B. 14
C. 15
D. 16
E. 17

11) A construction company is building a wall. The company can build 30 cm of the wall per minute. After 40 minutes construction, $\frac{3}{4}$ of the wall is completed. How high is the wall?

A. 9 m
B. 12 m
C. 16 m
D. 18 m
E. 20 m

12) 7 cubed is the same as:

A. 7×7
B. $7 \times 7 \times 7 \times 7$
C. 14
D. 343
E. 16807

13) When a number is subtracted from 24 and the difference is divided by that number, the result is 3. What is the value of the number?

A. 2
B. 4
C. 6
D. 12
E. 24

14) If car A drives 600 miles in 8 hours and car B drives the same distance in 7.5 hours, how many miles per hour does car B drive faster than car A?

A. 80
B. 75
C. 15
D. 10
E. 5

15) The ratio of boys to girls in a school is 2:3. If there are 600 students in a school, how many boys are in the school?

A. 540
B. 360
C. 300
D. 280
E. 240

16) $\frac{(7+5)^2}{4} + 5 = ?$

A. 41
B. 42
C. 43
D. 44
E. 45

17) If $y = 4ab + 3b^3$, what is y when $a = 2$ and $b = 3$?

A. 24
B. 31
C. 36
D. 51
E. 105

18) A company pays its employee $7000 plus 2% of all sales profit. If x is the number of all sales profit, which of the following represents the employee's revenue?

A. $0.02x$

B. $0.98x - 7000$

C. $0.02x + 7000$

D. $0.98x + 7000$

E. $0.09x$

19) Which of the following shows the numbers in increasing order?

A. $\frac{2}{3}, \frac{5}{7}, \frac{8}{11}, \frac{3}{4}$

B. $\frac{5}{7}, \frac{3}{4}, \frac{8}{11}, \frac{2}{3}$

C. $\frac{8}{11}, \frac{3}{4}, \frac{5}{7}, \frac{2}{3}$

D. $\frac{5}{7}, \frac{8}{11}, \frac{3}{4}, \frac{2}{3}$

E. None of the above

20) The area of a circle is 64 π. What is the circumference of the circle?

A. 8 π

B. 12 π

C. 16 π

D. 32 π

E. 64 π

21) If 60% of x equal to 30% of 20, then what is the value of $(x + 5)^2$?

A. 25.25
B. 26
C. 26.01
D. 225
E. 11,025

22) What is the greatest common factor of 36 and 54?

A. 28
B. 24
C. 18
D. 12
E. 8

23) The length of a rectangle is $\frac{5}{4}$ times its width. If the width is 16, what is the perimeter of this rectangle?

A. 36
B. 48
C. 72
D. 144
E. 180

24) Find $\frac{1}{4}$ of $\frac{2}{5}$ of 120?

A. 16

B. 12

C. 8

D. 4

E. 2

25) If the interior angles of a quadrilateral are in the ratio 1:2:3:4, what is the measure of the smallest angle?

A. 18°

B. 36°

C. 72°

D. 108°

E. 180°

IF YOU FINISH BEFORE TIME IS CALLED, YOU MAY CHECK YOUR WORK ON THIS SECTION ONLY. DO NOT TURN TO ANY OTHER SECTION IN THE TEST. **STOP**

SSAT UPPER LEVEL Math Practice Tests Answers and Explanations

SSAT Upper Level Mathematics Practice Test 1							
Section 1				Section 2			
1-	A	14-	A	1-	B	14-	E
2-	D	15-	D	2-	C	15-	D
3-	D	16-	B	3-	C	16-	B
4-	C	17-	B	4-	E	17-	D
5-	C	18-	B	5-	A	18-	E
6-	D	19-	B	6-	C	19-	E
7-	E	20-	C	7-	B	20-	A
8-	A	21-	B	8-	E	21-	B
9-	E	22-	D	9-	E	22-	E
10-	A	23-	D	10-	E	23-	C
11-	C	24-	B	11-	B	24-	C
12-	B	25-	D	12-	D	25-	D
13-	A			13-	D		

SSAT Upper Level Mathematics Practice Test 2							
Section 1				**Section 2**			
1-	**A**	14-	**A**	1-	**C**	14-	**E**
2-	**E**	15-	**C**	2-	**E**	15-	**E**
3-	**B**	16-	**A**	3-	**A**	16-	**A**
4-	**D**	17-	**B**	4-	**B**	17-	**E**
5-	**D**	18-	**D**	5-	**E**	18-	**C**
6-	**D**	19-	**C**	6-	**C**	19-	**A**
7-	**B**	20-	**A**	7-	**E**	20-	**C**
8-	**D**	21-	**A**	8-	**B**	21-	**D**
9-	**B**	22-	**A**	9-	**E**	22-	**C**
10-	**D**	23-	**C**	10-	**E**	23-	**C**
11-	**C**	24-	**D**	11-	**C**	24-	**B**
12-	**A**	25-	**D**	12-	**D**	25-	**B**
13-	**C**			13-	**C**		

Score Your Test

SSAT scores are broken down by its three sections: Verbal, Quantitative (or Math), and Reading. A sum of the three sections is also reported.

For the Upper Level SSAT, the score range is 500-800, the lowest possible score a student can earn is 500 and the highest score is 800 for each section. A student receives 1 point for every correct answer and loses $\frac{1}{4}$ point for each incorrect answer. No points are lost by skipping a question.

The total scaled score for an Upper Level SSAT is the sum of the scores for the quantitative, verbal, and reading sections. A student will also receive a percentile score of between 1-99% that compares that student's test scores with those of other test takers of same grade and gender from the past 3 years.

Use the following table to convert SSAT Upper level raw score to scaled score.

SSAT Upper Level Math Scaled Scores	
Raw Scores	**Scaled Score**
50	800
45	790
40	765
35	740
30	715
25	694
20	668
15	642
10	605
5	575
0	544
-5	512
- 10 and lower	500

SSAT Upper Level Math Practice Test 1

Section 1

1) Choice A is correct

Number of rotates in 12 second equals to: $\frac{300\times12}{8} = 450$

2) Choice D is correct

Number of packs equal to: $\frac{20}{3} \cong 6.677$

Therefore, the school must purchase 7 packs.

3) Choice D is correct

$\frac{25}{A} + 1 = 6 \rightarrow \frac{25}{A} = 6 - 1 = 5$

$\rightarrow 25 = 5A \rightarrow A = \frac{25}{5} = 5$

$25 + A = 25 + 5 = 30$

4) Choice C is correct

The area of the floor is: 6 cm × 24 cm = 144 cm^2

The number of tiles needed = 144 ÷ 8 = 18

5) Choice C is correct

A. Number of books sold in April is: 380

Number of books sold in July is: $760 \rightarrow \frac{380}{760} = \frac{38}{76} = \frac{1}{2}$

B. number of books sold in July is: 760

Half the number of books sold in May is: $\frac{1140}{2} = 570 \rightarrow 760 > 570$

C. number of books sold in June is: 190

Half the number of books sold in April is: $\frac{380}{2} = 190 \rightarrow 190 = 190$

D. $380 + 190 = 570 < 760$

E. $380 < 760$

6) Choice D is correct

$$12.124 \div 0.002 = \frac{\frac{12{,}124}{1{,}000}}{\frac{2}{1{,}000}} = \frac{12{,}124}{2} = 6{,}062$$

7) Choice E is correct

The digit in tens place is 1.

The digit in the thousandths place is 4.

Therefore; $1 + 4 = 5$

8) Choice A is correct

First, find the number.

Let x be the number. Write the equation and solve for x.

150 % of a number is 75, then:

$1.5 \times x = 75 \rightarrow x = 75 \div 1.5 = 50$

90 % of 50 is: $0.9 \times 50 = 45$

9) Choice E is correct

The amount of money that Jack earns for one hour: $\frac{\$616}{44} = \$14$

Number of additional hours that he needs to work in order to make enough money is:

$$\frac{\$826-\$616}{1.5\times\$14}=10$$

Number of total hours is: $44+10=54$

10) Choice A is correct

$$\frac{1\frac{3}{4}+\frac{1}{3}}{2\frac{1}{2}-\frac{15}{8}}=\frac{\frac{7}{4}+\frac{1}{3}}{\frac{5}{2}-\frac{15}{8}}=\frac{\frac{21+4}{12}}{\frac{20-15}{8}}=\frac{\frac{25}{12}}{\frac{5}{8}}=\frac{25\times8}{12\times5}=\frac{5\times2}{3\times1}=\frac{10}{3}\cong3.33$$

11) Choice C is correct

$1\leq x<4\rightarrow$ Multiply all sides of the inequality by 2. Then:

$2\times1\leq2\times x<2\times4\rightarrow2\leq2x<8$

All 1 to all sides. Then: $\rightarrow 2+1\leq2x+1<8+1\rightarrow\ 3\leq2x+1<8$

Minimum value of $2x+1$ is 3

12) Choice B is correct

$5\blacksquare11=\sqrt{5^2+11}=\sqrt{25+11}=\sqrt{36}=6$

13) Choice B is correct

Average = $\frac{\text{sum of terms}}{\text{number of terms}}$

The sum of the weight of all girls is: 18 × 60 = 1080 kg

The sum of the weight of all boys is: 32 × 62 = 1984 kg

The sum of the weight of all students is: 1080 + 1984 = 3064 kg

The average weight of the 50 students: $\frac{3064}{50}=61.28$

14) Choice A is correct

Let x be the capacity of one tank. Then, $\frac{2}{5}x = 200 \rightarrow x = \frac{200\times5}{2} = 500$ Liters

The amount of water in three tanks is equal to: $3 \times 500 = 1500$ Liters

15) Choice D is correct

$$7.5 \div 0.15 = \frac{7.5}{0.15} = \frac{\frac{75}{10}}{\frac{15}{100}} = \frac{75 \times 100}{15 \times 10} = \frac{75}{15} \times \frac{100}{10} = 5 \times 10 = 50$$

Average $= \frac{3064}{50} = 61.28$

16) Choice B is correct

Let x be the cost of one-kilogram orange, then: $3x + (2 \times 4.2) = 26.4 \rightarrow 3x + 8.4 = 26.4 \rightarrow 3x = 26.4 - 8.4 \rightarrow 3x = 18 \rightarrow x = \frac{18}{3} = \$6$

17) Choice B is correct

All angles in a triangle sum up to 180 degrees. Then:

$x = 20 + 125 = 145$

18) Choice B is correct

Let's review the choices provided.

A. 4. In 4 years, David will be 46 and Ava will be 10. 46 is not 4 times 10.
B. 6. In 6 years, David will be 48 and Ava will be 12. 48 is 4 times 12!
C. 8. In 8 years, David will be 80 and Ava will be 14. 50 is not 4 times 14.
D. 10. In 10 years, David will be 52 and Ava will be 16. 52 is not 4 times 16.
E. 14. In 14 years, David will be 56 and Ava will be 20. 56 is not 4 times 20.

19) Choice B is correct

Let b be the amount of time Alec can do the job, then,

$\frac{1}{a}+\frac{1}{b}=\frac{1}{100} \rightarrow \frac{1}{300}+\frac{1}{b}=\frac{1}{100} \rightarrow \frac{1}{b}=\frac{1}{100}-\frac{1}{300}=\frac{2}{300}=\frac{1}{150}$

Then: $b = 150$ minutes

20) Choice C is correct

The smallest number is -15. To find the largest possible value of one of the other five integers, we need to choose the smallest possible integers for four of them. Let x be the largest number. Then: $-70 = (-15) + (-14) + (-13) + (-12) + (-11) + x \rightarrow -70 = -65 + x$

$\rightarrow x = -70 + 65 = -5$

21) Choice B is correct

The equation of a line in slope intercept form is: $y = \text{m}x + b$

Solve for y.

$4x - 2y = 12 \Rightarrow -2y = 12 - 4x \Rightarrow y = (12 - 4x) \div (-2) \Rightarrow$

$y = 2x - 6$

The slope is 2. The slope of the line perpendicular to this line is:

$m_1 \times m_2 = -1 \Rightarrow 2 \times m_2 = -1 \Rightarrow m_2 = -\frac{1}{2}$

22) Choice D is correct

Use the information provided in the question to draw the shape.

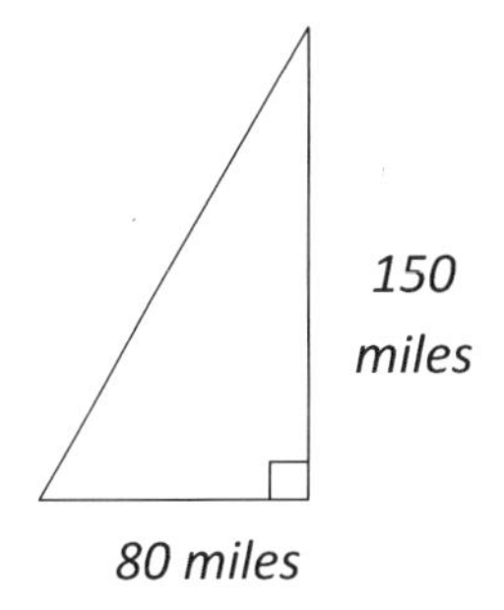

Use Pythagorean Theorem: $a^2 + b^2 = c^2$

$80^2 + 150^2 = c^2 \Rightarrow 6400 + 22500 = c^2 \Rightarrow 28900 = c^2 \Rightarrow c = 170$

23) Choice D is correct

The amount of money for x bookshelf is: $100x$

Then, the total cost of all bookshelves is equal to: $100x + 800$

The total cost, in dollar, per bookshelf is: $\frac{Total\ cost}{number\ of\ items} = \frac{100x+800}{x}$

24) Choice B is correct

Choices A, C and D are incorrect because 80% of each of the numbers is non-whole number.

A. 49, $80\%\ of\ 49 = 0.80 \times 49 = 39.2$

B. 35, $80\%\ of\ 35 = 0.80 \times 35 = 28$

C. 12, $80\%\ of\ 12 = 0.80 \times 12 = 9.6$

D. 32, $80\%\ of\ 32 = 0.80 \times 32 = 25.6$

E. 16, $80\%\ of\ 16 = 0.80 \times 16 = 12.8$

25) Choice D is correct

If the length of the box is 27, then the width of the box is one third of it, 9, and the height of the box is 3 (one third of the width). The volume of the box is:

V = (length)(width)(height) = (27) (9) (3) = 729

SSAT UPPER LEVEL Math Practice Test 1

Section 2

1) Choice B is correct

$\frac{17+11}{2} = \frac{28}{2} = 14$ Then, $14 - 11 = 3$

2) Choice C is correct

Let's review the choices provided:

A. $x = 2 \rightarrow$ The perimeter of the figure is: $2 + 4 + 2 + 2 + 2 = 12 \neq 20$

B. $x = 3 \rightarrow$ The perimeter of the figure is: $2 + 4 + 2 + 3 + 3 = 14 \neq 20$

C. $x = 6 \rightarrow$ The perimeter of the figure is: $2 + 4 + 2 + 6 + 6 = 20 = 20$

D. $x = 9 \rightarrow$ The perimeter of the figure is: $2 + 4 + 2 + 9 + 9 = 26 \neq 20$

E. $x = 12 \rightarrow$ The perimeter of the figure is: $2 + 4 + 2 + 12 + 12 = 32 \neq 20$

3) Choice C is correct

$\frac{91501}{305} \cong 300.0032 \cong 300$

4) Choice E is correct

Alex's mark is k less than Jason's mark. Then, from the choices provided Alex's mark can only be $16 - k$.

5) Choice A is correct

The Area that one liter of paint is required: 72cm × 100cm = 7,200cm^2

Remember: 1 m^2 = 10,000 cm^2 (100 × 100 = 10,000), then, 7,200cm^2 = 0.72 m^2

Number of liters of paint we need: $\frac{36}{0.72} = 50$ liters

6) Choice C is correct

Let x be the original price. If the price of the sofa is decreased by 25% to \$420, then:

$75\ \%$ of $x = 420 \Rightarrow 0.75x = 420 \Rightarrow x = 420 \div 0.75 = 560$

7) Choice B is correct

$$750 - 7\frac{7}{15} = (749 - 7) + \left(\frac{15}{15} - \frac{7}{15}\right) = 742\frac{8}{15}$$

8) Choice E is correct

Number of times that the driver rests $= \frac{20}{4} = 5$

Driver's rest time $=$ 1 hour and 12 minutes $=$ 72 minutes

Then, 5×72 minutes $=$ 360 minutes

1 hour $=$ 60 minutes $\rightarrow$ 360 minutes $=$ 6 hours

9) Choice E is correct

Let's review the options provided:

A. $10 \times \frac{1}{2} = \frac{10}{2} = 5 = 5$
B. $25 \times \frac{1}{5} = \frac{25}{5} = 5 = 5$
C. $2 \times \frac{5}{2} = \frac{10}{2} = 5 = 5$
D. $6 \times \frac{5}{6} = \frac{30}{6} = 5 = 5$
E. $5 \times \frac{1}{5} = \frac{5}{5} = 1 \neq 5$

10) Choice E is correct

Find the difference of each pairs of numbers:

2, 3, 5, 8, 12, 17, 23, ___, 38

The difference of 2 and 3 is 1, 3 and 5 is 2, 5 and 8 is 3, 8 and 12 is 4, 12 and 17 is 5, 17 and 23 is 6, 23 and next number should be 7. The number is 23 + 7 = 30

11) Choice B is correct

Number of Mathematics book: $0.3 \times 840 = 252$

Number of English books: $0.15 \times 840 = 126$

Product of number of Mathematics and number of English books: $252 \times 126 = 31{,}752$

12) Choice D is correct

The angle α is: $0.3 \times 360 = 108°$

The angle β is: $0.15 \times 360 = 54°$

13) Choice D is correct

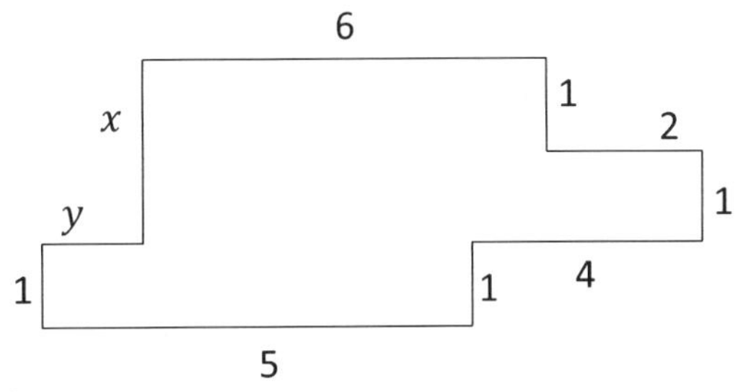

6, x, 1, 2, y, 1, 1, 1, 4, 5

$x + 1 = 1 + 1 + 1 \rightarrow x = 2$

$y + 6 + 2 = 5 + 4 \rightarrow y + 8 = 9 \rightarrow y = 1$

Then, the perimeter is:

$1 + 5 + 1 + 4 + 1 + 2 + 1 + 6 + 2 + 1 = 24$

14) Choice E is correct

$3y + 5 < 29 \rightarrow 3y < 29 - 5 \rightarrow 3y < 24 \rightarrow y < 8$

The only choice that is less than 8 is E.

15) Choice D is correct

The capacity of a red box is 20% bigger than the capacity of a blue box and it can hold 30 books. Therefore, we want to find a number that 20% bigger than that number is 30. Let x be that number. Then: $1.20 \times x = 30$, Divide both sides of the equation by 1.2. Then:

$$x = \frac{30}{1.20} = 25$$

16) Choice B is correct

Since, E is the midpoint of AB, then the area of all triangles DAE, DEF, CFE and CBE are equal.

Let x be the area of one of the triangle, then: $4x = 100 \rightarrow x = 25$

The area of DEC $= 2x = 2(25) = 50$

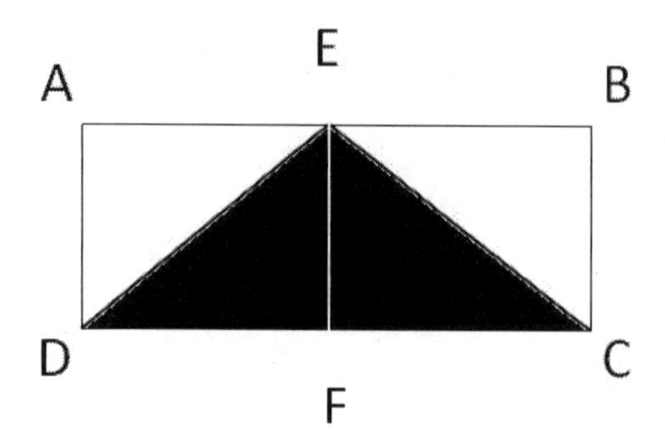

17) Choice D is correct

Amount of available petrol in tank: $50.2 - 5.28 - 25.9 + 10.31 = 28.79$ liters

18) Choice E is correct

We have two equations and three unknown variables, therefore x cannot be obtained.

19) Choice E is correct

Let put some values for a and b. If $a = 9$ and $b = 2 \rightarrow a \times b = 18 \rightarrow \frac{18}{3} = 6 \rightarrow 18$ is divisible by 3 then;

A. $a + b = 9 + 2 = 11$ is not divisible by 3

B. $3a - b = (3 \times 9) - 2 = 27 - 2 = 25$ is not divisible by 3

If $a = 11$ and $b = 3 \rightarrow a \times b = 33 \rightarrow \frac{33}{3} = 11$ is divisible by 3 then;

C. $a - 3b = 11 - (3 \times 3) = 11 - 9 = 2$ is not divisible by 3

D. $\frac{a}{b} = \frac{11}{3}$ is not divisible by 3

E. $4 \times 11 \times 3 = 132$

132 is divisible by 3. If you try any other numbers for a and b, you will get the same result.

20) Choice A is correct

Perimeter of figure A is: $2\pi r = 2\pi \frac{16}{2} = 16\pi = 16 \times 3 = 48$

Area of figure B is: $5 \times 12 = 60$

Average $= \frac{48+60}{2} = \frac{108}{2} = 54$

21) Choice B is correct

The angles on a straight line add up to 180 degrees. Let's review the choices provided:

A. $y = 15 \rightarrow x + 25 + y + 2x + y = 35 + 25 + 15 + 2(35) + 15 = 160 \neq 180$

B. $y = 25 \rightarrow x + 25 + y + 2x + y = 35 + 25 + 25 + 2(35) + 25 = 180$

C. $y = 35 \rightarrow x + 25 + y + 2x + y = 35 + 25 + 35 + 2(35) + 35 = 200 \neq 180$

D. $y = 45 \rightarrow x + 25 + y + 2x + y = 35 + 25 + 45 + 2(35) + 45 = 220 \neq 180$

E. $y = 55 \rightarrow x + 25 + y + 2x + y = 35 + 25 + 55 + 2(35) + 55 = 240 \neq 180$

22) Choice E is correct

Let's review the choices provided:

A. $x = \frac{1}{2} \rightarrow \frac{5}{9} + \frac{1}{2} = \frac{10+9}{18} = \frac{19}{18} \cong 1.056 < 2$

B. $x = \frac{3}{5} \rightarrow \frac{5}{9} + \frac{3}{5} = \frac{25+27}{45} = \frac{52}{45} \cong 1.16 < 2$

C. $x = \frac{4}{5} \rightarrow \frac{5}{9} + \frac{4}{5} = \frac{25+36}{45} = \frac{61}{45} \cong 1.36 < 2$

D. $x = \frac{4}{3} \rightarrow \frac{5}{9} + \frac{4}{3} = \frac{5+12}{9} = \frac{17}{9} \cong 1.89 < 2$

E. $x = \frac{5}{3} \rightarrow \frac{5}{9} + \frac{5}{3} = \frac{5+15}{9} = \frac{20}{9} \cong 2.2 > 2$

Only choice be is correct.

23) Choice C is correct

Set of numbers that are not composite between 1 and 25: A= $\{2, 3, 5, 7, 11, 13, 17, 19, 23\}$

$$\text{Probability} = \frac{\textit{number of desired outcomes}}{\text{number of total outcomes}} = \frac{9}{25}$$

24) Choice C is correct

$$712 \div 3 = \frac{712}{3} = \frac{700 + 10 + 2}{3} = \frac{700}{3} + \frac{10}{3} + \frac{2}{3}$$

25) Choice D is correct

$$\frac{2}{5} \times 25 = \frac{50}{5} = 10$$

SSAT UPPER LEVEL Math Practice Tests Explanations

SSAT UPPER LEVEL Math Practice Test 2

Section 1

1) Choice C is correct

$0.03 \times 12.00 = \frac{3}{100} \times \frac{12}{1} = \frac{36}{100} = 0.36$

2) Choice E is correct

$x - 10 = -10 \rightarrow x = -10 + 10 \rightarrow x = 0$, Then; $x \times 3 = 0 \times 3 = 0$

3) Choice B is correct

Digit 4 is in the tenths place.

4) Choice D is correct

Number of packs needed equals to: $\frac{16}{3} \cong 5.33$

Then Mia must purchase 6 packs.

5) Choice D is correct

His average speed was: $\frac{2.5}{0.5} = 5$ miles per hour

6) Choice D is correct

The amount of flour is: $x - 55$

7) Choice B is correct

The time it takes to drive from city A to city B is: $\frac{2600}{68} = 38.23$

It's approximately 38 hours.

8) Choice B is correct

The perimeter of the quadrilateral is: $6 + 20 + 9 + 31 = 66$

9) Choice C is correct

Number of males in the classroom is: $60 - 42 = 18$

Then, the percentage of males in the classroom is: $\frac{18}{60} \times 100 = 0.3 \times 100 = 30\%$

10) Choice D is correct

Let x be the fourth quarter rate, then: $\frac{92+38+86+x}{4} = 90$

Multiply both sides of the above equation by 4. Then:

$$4 \times \left(\frac{92 + 38 + 86 + x}{4}\right) = 4 \times 90 \rightarrow 92 + 88 + 86 + x = 360 \rightarrow 266 + x = 360 \rightarrow x = 360 - 266 = 94$$

11) Choice C is correct

Let x be the number. Write the equation and solve for x.

$\frac{2}{3} \times 18 = \frac{2}{5}x \rightarrow \frac{2 \times 18}{3} = \frac{2x}{5}$, use cross multiplication to solve for x.

$5 \times 36 = 2x \times 3 \Rightarrow 180 = 6x \Rightarrow x = 30$

12) Choice A is correct

For one person the total cost is: $\$8.25$

Therefore, for four persons, the total cost is: $4 \times \$8.25 = \$33$

13) Choice C is correct

Three times of 24,000 is 72,000. One sixth of them cancelled their tickets.

One sixth of 72,000 equals 12,000 (1/6 × 72000 = 12000).

60,000 (72000 – 12000 = 60000) fans are attending this week

14) Choice A is correct

The equation of a line in slope intercept form is: $y = \mathrm{m}x + b$

Solve for y.

$7x + y = 12 \Rightarrow y = -7x + 12$

$y = -7x + 12$

The slope of this line is -7.

The slope of the line perpendicular to this line is:

$m_1 \times m_2 = -1 \Rightarrow -7 \times m_2 = -1 \Rightarrow m_2 = \frac{1}{7}$

15) Choice C is correct

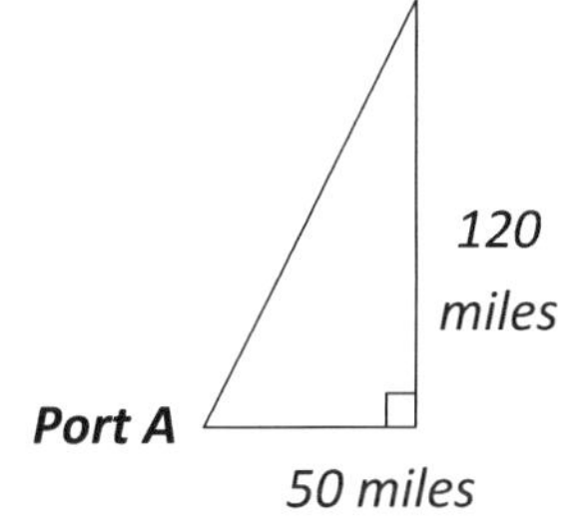

Use the information provided in the question to draw the shape.

Use Pythagorean Theorem: $a^2 + b^2 = c^2$

$50^2 + 120^2 = c^2 \Rightarrow 2500 + 14400 = c^2 \Rightarrow 16900 = c^2 \Rightarrow c = 130$

16) Choice A is correct

Let x be the number of years. Therefore, \$2,000 per year equals $2000x$.

Starting from $24,000 annual salary means you should add that amount to $2000x$.

Income more than that is:

I > $2000\,x\ +\ 24000$

17) Choice B is correct

Ratio of women to men in city A: $\frac{570}{600}$=0.95

Ratio of women to men in city B: $\frac{291}{300}$=0.97

Ratio of women to men in city C: $\frac{665}{700}$=0.95

Ratio of women to men in city D: $\frac{528}{550}$=0.96

0.97 is the maximum ratio of woman to man in the four cities.

18) Choice D is correct

Percentage of men in city A $= \frac{600}{1170} \times 100 = 51.28\%$

Percentage of women in city C $= \frac{665}{1365} \times 100 = 48.72\%$

Percentage of men in city A to percentage of women in city C $= \frac{51.28}{48.72} = 1.05$

19) Choice C is correct

Let the number of women should be added to city D be x, then:

$$\frac{528 + x}{550} = 1.2 \rightarrow 528 + x = 550 \times 1.2 = 660 \rightarrow x = 132$$

20) Choice A is correct

The length of MN is equal to: $3x + 5x = 8x$

Then: $8x = 40 \rightarrow x = \frac{40}{8} = 5$

The length of ON is equal to: $5x = 5 \times 5 = 25$ cm

21) Choice A is correct

$$\frac{x}{2+3} = \frac{y}{10-7} \rightarrow \frac{x}{5} = \frac{y}{3} \rightarrow 5y = 3x \rightarrow y = \frac{3}{5}x$$

22) Choice A is correct

$$\sqrt[5]{x^{16}} = \sqrt[5]{x^{15} \times x} = \sqrt[5]{x^{15}} \times \sqrt[5]{x} = x^{\frac{15}{5}} \times \sqrt[5]{x} = x^3\sqrt[5]{x}$$

23) Choice C is correct

The average speed of John is: 150 ÷ 6 = 25 km

The average speed of Alice is: 180 ÷ 4 = 45 km

Write the ratio and simplify.

25 : 45 ⇒ 5 : 9

24) Choice D is correct

Smallest 4–digit number is 1000, and biggest 4–digit number is 9999. The difference is: 8999

25) Choice D is correct

Perimeter of rectangle A is equal to: $2 \times (8+4) = 2 \times 12 = 24$

Perimeter of rectangle A is equal to: $2 \times (5+4) = 2 \times 9 = 18$

Therefore: $\frac{18}{24} \times 100 = 0.75 \times 100 = 75\%$

SSAT UPPER LEVEL Math Practice Test 2

Section 2

1) Choice C is correct

$$5\frac{3}{7}\times 4\frac{1}{5}=\frac{38}{7}\times\frac{21}{5}=\frac{38\times 21}{7\times 5}=\frac{798}{35}=\frac{114}{5}=22\frac{4}{5}$$

2) Choice E is correct

A. $\frac{2}{3}\times\frac{9}{5}=\frac{6}{5}$ is not equal to whole number

B. $\frac{1}{2}+\frac{1}{4}=\frac{2+1}{4}=\frac{3}{4}$ is not equal to whole number

C. $\frac{21}{6}=\frac{7}{2}=3.5$ is not equal to whole number

D. $2.5+1=3.5$ is not equal to whole number

E. $2.5+\frac{7}{2}=2.5+3.5=6$ is a whole number

3) Choice A is correct

$$0.42\times 11.8=\frac{42}{100}\times\frac{118}{10}=\frac{42\times 118}{100\times 10}=\frac{4956}{1000}=4.956$$

4) Choice B is correct

The question is this: 530.40 is what percent of 624?

Use percent formula:

$$\text{Part}=\frac{percent}{100}\times\text{whole}$$

$530.40 = \frac{\text{percent}}{100} \times 624 \rightarrow 530.40 = \frac{\text{percent} \times 624}{100} \rightarrow 53040 = \text{percent} \times 624$

Then, Percent $= \frac{53040}{624} = 85$

530.40 is 85 % of 624. Therefore, the discount is: 100% – 85% = 15%

5) Choice E is correct

Let x be the number, then; $\frac{2}{5}x = 12 \rightarrow x = \frac{5 \times 12}{2} = 30$

Therefore: $\frac{2}{3}x = \frac{2}{3} \times 30 = 20$

6) Choice C is correct

Use distance formula:

Distance = Rate × time ⇒ 420 = 50 × T, divide both sides by 50. 420 / 50 = T ⇒ T = 8.4 hours.

Change hours to minutes for the decimal part. 0.4 hours = 0.4 × 60 = 24 minutes

7) Choice E is correct

Use formula of rectangle prism volume.

V = (length) (width) (height) ⇒ 2000 = (25) (10) (height) ⇒ height = 2000 ÷ 250 = 8

8) Choice B is correct

Use simple interest formula:

$I = prt$

(I = interest, p = principal, r = rate, t = time)

$I = (12000)(0.035)(2) = \$840$

9) Choice E is correct

The angle x and 35 are complementary angles. Therefore:

$x + 35 = 180$

$180° - 35° = 145°$

10) Choice E is correct

Number of boxes equal to: $\frac{408}{24} = \frac{102}{6} = \frac{34}{2} = 17$

11) Choice C is correct

The rate of construction company $= \frac{30 \text{ cm}}{1 \text{ min}} = 30$ cm/min

Height of the wall after 40 min$= \frac{30 \text{ cm}}{1 \text{ min}} \times 40 \text{ min} = 1200 \text{ cm}$

Let x be the height of wall, then $\frac{3}{4}x = 1200 \text{ cm} \rightarrow x = \frac{4 \times 1200}{3} \rightarrow x = 1600 \text{ cm} = 16$ m

12) Choice D is correct

7 cubed is: $7 \times 7 \times 7 = 49 \times 7 = 343$

13) Choice C is correct

Let's review the choices provided:

A. $24 - 2 = 22 \rightarrow \frac{22}{2} = 11 \neq 3$

B. $24 - 4 = 20 \rightarrow \frac{20}{4} = 5 \neq 3$

C. $24 - 6 = 18 \rightarrow \frac{18}{6} = 3 = 3$

D. $24 - 12 = 12 \rightarrow \frac{12}{12} = 1 \neq 3$

E. $24 - 24 = 0 \rightarrow \frac{0}{24} = 0 \neq 3$

14) Choice E is correct

Speed of car A is: $\frac{600}{8} = 75$ Km/h

Speed of car B is: $\frac{600}{7.5} = 80$ Km/h

$\rightarrow 80 - 75 = 5$ Km/h

15) Choice E is correct

The ratio of boys to girls is 2:3. Therefore, there are 2 boys out of 5 students. To find the answer, first divide the total number of students by 5, then multiply the result by 2.

$600 \div 5 = 120 \Rightarrow 120 \times 2 = 240$

16) Choice A is correct

$$\frac{(7+5)^2}{4} + 5 = \frac{(12)^2}{4} + 5 = \frac{144}{4} + 5 = 36 + 5 = 41$$

17) Choice E is correct

$y = 4ab + 3b^3$

Plug in the values of a and b in the equation: $a = 2$ and $b = 3$

$y = 4\,(2)\,(3) + 3\,(3)^3 = 24 + 3(27) = 24 + 81 = 105$

18) Choice C is correct

x is the number of all sales profit and 2% of it is:

$2\% \times x = 0.02x$

Employee's revenue: $0.02x + 7000$

19) Choice A is correct

$\frac{2}{3} \cong 0.67$ $\frac{5}{7} \cong 0.71$ $\frac{8}{11} \cong 0.73$ $\frac{3}{4} = 0.75$

20) Choice C is correct

Use the formula of the area of circles.

Area $= \pi r^2 \Rightarrow 64\,\pi = \pi r^2 \Rightarrow 64 = r^2 \Rightarrow r = 8$

Radius of the circle is 8. Now, use the circumference formula:

Circumference = $2\pi r = 2\pi\,(8) = 16\,\pi$

21) Choice D is correct

$0.6x = (0.3) \times 20 \rightarrow x = 10 \rightarrow (x+5)^2 = (10+5)^2 = (15)^2 = 225$

22) Choice C is correct

Prime factorizing of 36= $2 \times 2 \times 3 \times 3$

Prime factorizing of 54= $2 \times 3 \times 3 \times 3$

GCF= $2 \times 3 \times 3 = 18$

23) Choice C is correct

Length of the rectangle is: $\frac{5}{4} \times 16 = 20$

Perimeter of rectangle is: $2 \times (20 + 16) = 72$

24) Choice B is correct

$\frac{2}{5}$ Of $120 = \frac{2}{5} \times 120 = 48$

$\frac{1}{4}$ Of $48 = \frac{1}{4} \times 48 = 12$

25) Choice B is correct

The sum of all angles in a quadrilateral is 360 degrees.

Let x be the smallest angle in the quadrilateral. Then the angles are:

$x, 2x, 3x, 4x$

$x + 2x + 3x + 4x = 360 \rightarrow 10x = 360 \rightarrow x = 36$

The angles in the quadrilateral are: 36°, 72°, 72°, and 180°

The smallest angle is 36 degrees.

"Effortless Math" Publications

Effortless Math authors' team strives to prepare and publish the best quality Mathematics learning resources to make learning Math easier for all. We hope that our publications help you or your student Math in an effective way.

We all in Effortless Math wish you good luck and successful studies!

Effortless Math Authors